DIVINE WIND

▲

Impending. Oil painting by American artist Amy Marx, 2002.

T he Voice of the Sea is never one voice, but a tumult of many voices—voices of drowned men,—the mutterings of multitudinous dead,—the moaning of innumerable ghosts, all rising, to rage against the living, at the great Witch-call of storms....
—Lafcadio Hearn (1850–1904), *Chita: A Memory of Last Island*

DIVINE WIND

The History and Science of Hurricanes

Kerry Emanuel

OXFORD

UNIVERSITY PRESS

2005

OXFORD

UNIVERSITY PRESS

Oxford University Press, Inc., publishes works that further
Oxford University's objective of excellence
in research, scholarship, and education.

Oxford New York
Auckland Cape Town Dar es Salaam Hong Kong Karachi
Kuala Lumpur Madrid Melbourne Mexico City Nairobi
New Delhi Shanghai Taipei Toronto

With offices in
Argentina Austria Brazil Chile Czech Republic France Greece
Guatemala Hungary Italy Japan Poland Portugal Singapore
South Korea Switzerland Thailand Turkey Ukraine Vietnam

Published by Oxford University Press, Inc.
198 Madison Avenue, New York, NY 10016
www.oup.com

Designed and typeset by Scott and Emily Santoro, Worksight

Library of Congress Cataloging-in-Publication Data
Emanuel, Kerry A., 1955–
  Divine wind : the history and science of hurricanes / Kerry Emanuel.
  p. cm. Includes bibliographical references and index.
ISBN-13: 978-0-19-514941-8
ISBN-10: 0-19-514941-6

1. Hurricanes--History.
I. Title.
QC944.E43 2005
551.55'2'09--dc22
2004013078

1 3 5 7 9 8 6 4 2

Printed in China on acid-free paper

Contents

List of Illustrated Passages

Preface

Of all the natural phenomena that affect our planet, the hurricane is among the most deadly and destructive. Hurricanes have killed more people worldwide in the last fifty years than any other natural cataclysm, and a single storm—Andrew of 1992—was the most expensive natural disaster in history. At the same time, the magnificence of these great tempests has influenced artists, writers, and naturalists through all time, inspiring works from ancient Mayan hieroglyphs to abstract renditions of recent storms, from the plays of William Shakespeare to a novel by Joseph Conrad. The terror and awe of rotating windstorms are conveyed in all of the major religious texts: the Bible, for example, contains numerous references to whirlwinds, invariably invoked as instruments of God. Hurricanes have altered both natural and political geography, cutting new inlets with the same ease with which they dispatch entire navies. As destructive of human ends as they may be, they nonetheless play a vital role in certain tropical ecosystems and may, ironically, prove essential to the comparative stability of tropical climate.

It is my ambition to portray the hurricane as it is seen from the perspectives of history, art, and science, so as to form an integrated appreciation of the phenomenon. To help accomplish this, the chapters of this book alternate between accounts of hurricanes that have played important roles in history and a scientific narrative that describes some of what has been learned about the nature of these storms. Interspersed throughout are works of art and literature that help convey the human experience of hurricanes. Some (but I suspect very few) of my scientific colleagues may complain that the sentiments expressed by the art and literature distract from what ought to be an impersonal pursuit of scientific truth. I remind them that there is not a single one of us who is not driven forward by a passion for our work, and there is no shame in admitting that. To those who feel that science need

◄

Hurricane Elena over the Gulf of Mexico in September, 1985, photographed by a crew member onboard the space shuttle Discovery.

play no role in the appreciation of nature, who feel that scientists are motivated primarily by the desire to demystify and to exercise power over nature, I ask you to remember how you felt about science when you were in third grade, when knowledge stimulated rather than extinguished wonder, and to contemplate the words of Albert Einstein, that the most beautiful experience we can have is that of the mysterious. I believe that most scientists are motivated more by the desire to find new mysteries than to solve known problems.

To one accustomed to publishing in scholarly journals, it becomes instinctive to cite all sources that were brought to bear on one's own contribution, yet to do so would considerably compromise the tone and flow of a book like this. I am left with the impossible task of summarizing in a few brief sentences the vast quantity of inspiration and help I have received from many others. I begin with two books—now quite dated—that served as models for this one. In her magnificent *Hurricane*, Marjory Stoneman Douglas presents a riveting account of the effect that tropical cyclones had on exploration and settlement of the New World, interwoven with discussion of what was then known about the physics of hurricanes. I relied on this work for much of the historical material presented here. In his Time-Life series book *Weather*, Philip Thompson showed me that historical, artistic, and scientific material can be skillfully interwoven to illustrate a complex topic in an entertaining and informative way. Other sources are given in the reference list at the back of the book. I am indebted to many for providing source material, photographs, and advice. These include Lixion Avila, Mike Black, Howie Bluestein, Ray Hennessey, Kam-Biu Liu, Frank Marks, Peter Perdue, Jim Pritchet, and Joel Sloman. This book would not have been undertaken without the encouragement of my editor, Joyce Berry of Oxford University Press. In addition to editing often unreadable drafts, she worked tirelessly to obtain many needed permissions, artwork, and works of literature while patiently enduring a steady stream of polemics about publishing. Work on this project made a virtual widow of my wife, Susan, and orphan of my son, David. To them I dedicate this book.

—Kerry Emanuel, Lexington, April 19, 2004

Supplementary material for Divine Wind and more
information about hurricanes may be found at:
http://www.oup.com/us/divinewind

▲

The Coming Storm.
Watercolor by American
artist Winslow Homer,
1901.

The summer's sun was sinking down 'neath Binion's waveless bay,
And burnishing its rippling tide with many a golden ray;
The zephyrs stayed their wanton steps, and hushed their every breath—
The scene was still, the bay was bright and undisturbed as death.
Tall Raghlin looked with queenly pride far out into the main,
And Binion threw its giant shade across the watery plain;
And fair Donaff in distance blue raised up its head on high,
And caught the sun's expiring beams, and kissed the cloudless sky—
No dark spot dimm'd the broad expanse that spann'd the silent sea;
That was fair as eastern bride, and brighter far than she!

A tiny boat, like speck of snow, on ocean's bosom hoar,
Had spread its sails at early morn, and left that lonely shore;
The noontide sun had seen it far out on the watery track,
And vesper lit her dazzling lamp to guide the wanderer back.
The idle sails now flap the mast, no breeze disturbs the sea,
And through Lagg Bar the angry tide for once steals silently;
The boatmen press the pliant oars, and raise the jocund song—
They pass the tower of Malin Head girt round by barriers strong;

And Tullagh's strand is full in view, and seen is rough Maymore—

Full well these boatmen know each spot from Doagh to Leenan Shore!

But just athwart the day-god's track a sudden gloom has passed,

As if the night her sombre veil across the day had cast;

A vivid flash lights up that gloom, the sudden thunder rolls—

It peals along the startled heavens, and roars around the poles.

The gushing rain comes dancing forth in drenching torrents wild,

And leaps the whirlwind from its throne of storms on storm-clouds piled,

It sweeps the main with tyrant might—upheaves the tranquil bay—

And dashes o'er the troubled sea like dolphin at its play;

It crests the wave with snowy foam, throws billows mountain high,

And rears up watery spires that pierce the bosom of the sky!

The storm has ceased, the night is on, and sighs the dying gale,

And quick the swollen streamlets run in murmurs down the vale;

And where's the boat—poor tiny thing—that rode the waves at morn,

And spread in pride its snowy sail like butterfly just born?

And where's the crew that mann'd that boat—Clonmany's seamen bold—

Who feared no tide, disdained all storms, felt not the winter's cold?

They've sunk beneath the billow's breast, down in the salt-sea wave—

No humble cross in hallowed spot shall mark their lonely grave.

Their shroud shall be the sea-weed green, their tomb the ocean sand,

Their epitaph—the tale which tells their fate upon the land.

The summer's sun again looks down on Binion's waveless bay,

And sees no trace which marks the storm that swept it yesterday;

But there are hearts beneath that tide cold, cold as winter's snow,

Which never more shall feel life's joys, nor taste its cup of woe.

On yester-morn those hearts were glad—their life blood bounded free—

The tempest swept across the deep, and sunk them in the sea!

There shall they sleep regarding not the storms that o'er them rave;

No sound of busy life shall break the stillness of their grave;

Eternal hurricanes may roll unheeded o'er their head—

No voice they'll hear but that which cries:—'Arise, arise ye dead!'

    —J. K. O'Doherty (1833–1907), "The Hurricane"

1   Kamikaze

Behold, a whirlwind of the Lord is gone forth in fury, even a grievous whirlwind: it shall fall grievously upon the head of the wicked.
    —Jeremiah 23:19

Were it not for two typhoons, Japan might be part of China today.

In the year 1259, Kublai Khan, the grandson of Genghis Khan, became emperor of Mongolia and renamed it Yuan, meaning "first beginning," reflecting his aspirations for the empire. In 1230, the Mongols had conquered northern China, and between 1231 and 1238, they overran the Korean peninsula. As the Mongols took it as their mission to conquer as much land as possible, Japan, separated by only 150 km (100 mi) from Korea, rightly feared invasion. Between 1267 and 1274, Kublai Khan sent a number of emissaries to Japan demanding that its emperor submit or face an invasion. These emissaries were usually turned back before their message reached the emperor; in any event, the Mongols never received a response.

Thus it came to pass that Kublai mounted an invasion to conquer Japan. Although the Mongols knew nothing of seafaring, the newly conquered Koreans were expert shipbuilders and navigators and were put to work assembling an impressive naval fleet.

On October 29, 1274, the invasion began. Some 40,000 men, including about 25,000 Mongolians and Chinese, 8,000 Korean troops, and 7,000 Chinese and Korean seamen, set sail from Korea in about 900 ships. At first, the expedition went well for the Mongols. After quickly overrunning several small islands off the northwest coast of Kyūshū, the force landed at the harbors of Imazu and Hakata in Kyūshū on November 19. The Japanese defenders were horrified by the Mongol cavalry charging off the beaches, steeped as they were in the tradition of hand-to-hand combat between knightly warriors. With fewer troops and inferior weapons, the Japanese were rapidly pushed back into the interior. But at nightfall, the Korean pilots sensed an approaching storm and begged their reluctant Mongol commanders to put the invasion force back to sea lest it be trapped on the coast

Figure 1.1 Scene from the thirteenth-century Mongol invasion scrolls, based on a narrative written by the Japanese warrior Takezaki Suenaga.

▲

and its ships destroyed at anchor. The next morning, the Japanese, who had wished for nothing more than a delay in the Mongol assault to allow reinforcements to arrive, were surprised and delighted to see the last of the Mongol armada struggling to regain the open ocean in the midst of a great storm. The ships of the time were no match for the tempest, and many foundered or were dashed to bits on the rocky coast. Nearly 13,000 men perished, mostly by drowning. The Mongols had been routed by a typhoon.

Even as Kublai Khan was mounting his Japanese offensive, he was waging a bitter war of conquest against southern China, whose people had resisted him for 40 years. But finally, in 1279, the last of the southern provinces, Canton, fell to the Mongol forces, and China was united under one ruler for the first time in three hundred years. Buoyed by success, Kublai again tried to bully Japan into submission. But this time the Japanese executed his emissaries, enraging him and thereby paving the way for a second invasion. Knowing this was inevitable, the Japanese went to work building coastal fortifications, including a massive dike around Hakozaki Bay, which encompasses the site of the first invasion.

The second Mongol invasion of Japan assumed staggering proportions. One armada consisting of 40,000 Mongols, Koreans, and north Chinese was to sail from Korea, while a second, larger force of some 100,000 men was to set out from various ports in south China. To gauge the size of this expeditionary force, consider that the Norman conquest of Britain in 1066 engaged 5,000 men.

The invasion plan called for the two armadas to join forces in the spring, before the summer typhoon season, but the southern force was late, delaying the invasion until late June 1281. When they arrived along the coast of Kyūshū, they met fierce resistance from Japan's samurai warriors, who were better prepared and now knew what to expect from the Mongols. The defenders held back the invading forces for six weeks, until, on the fifteenth and sixteenth of August, history repeated itself. Once again, the Korean and south Chinese mariners sensed the approach of a typhoon and attempted to put to sea. But the fleet was so unwieldy and poorly coordinated that many of the ships collided at the entrance of Imari Bay and were smashed by the typhoon, as were most of those that made it to the open ocean. Kublai managed to escape on an undamaged ship, but left his men to die at the hands of the storm and the samurai. The wreckage and loss of life was staggering. Once again, Kublai Khan's designs on Japan were defeated by a typhoon, and never again did he attempt such an invasion.

As a direct result of these famous routs, the Japanese came to think of the typhoon as a "divine wind," or *kamikaze*, sent by their gods to deliver their land from invaders.

Seven hundred years later, in 1981, Torao Mozai, a Tokyo University engineering professor, conducted sonar-aided digs in Japan's Imari Bay and turned up numerous artifacts from the Mongol invasions, including spearheads, war helmets, a cavalry officer's iron sword, and stone anchor stocks. Fisherman came forward with other objects they had pulled up in their nets. Excavations continue to this day and have yielded many more treasures from these ancient invasions, including ceramic pots and ship anchors.

Six hundred and sixty-three years after Kublai Khan's last defeat on the shores of Japan, another naval commander, facing the same enemy, made the same mistake twice in succession. Aircraft, weather maps, and steel ships were not enough to prevent Admiral William F. "Bull" Halsey, commander of the U.S. Third Fleet in World War II, from meeting disaster in the form of two typhoons. We shall return to Admiral Halsey's plight in Chapter 23.

▲

Storm II The Hurricane.
Acrylic painting by
American artist Suzette
Barton Chandler.

L o, Lord, Thou ridest!
Lord, Lord, Thy swifting heart

Naught stayeth, naught now bideth
But's smithereened apart!

Ay! Scripture flee'th stone!
Milk-bright, Thy chisel wind

Rescindeth flesh from bone
To quivering whittlings thinned—

Swept-whistling straw! Battered,
Lord, e'en boulders now out-leap

Rock sockets, levin-lathered!
Nor, Lord, may worm out-creep

Thy drum's gambade, its plunge abscond!
Lord God, while summits crashing

Whip sea-kelp screaming on blond
Sky-seethe, high heaven dashing—

Thou ridest to the door, Lord!
Thou bidest wall nor floor, Lord!

      —Hart Crane (1899–1932), "The Hurricane"

2 Anatomy of a Meteorological Monster

The storm was in the form of a great whirlwind.
—William Redfield, 1831

The early settlers of the New World were adept at describing the effects of hurricanes, but they were not able to form any conception of their structure. It was not until the early nineteenth century that naturalists began to accept the idea that hurricanes are vast vortices, or whirlwinds. In 1821, William Redfield, a New England saddler and amateur meteorologist, traveled through Connecticut to inspect the damage wrought by a hurricane that had raced northward through the mid-Atlantic states and New England. He noticed that although trees had fallen toward the northwest in the eastern part of the state, they pointed southeastward in the western part. He then hypothesized that hurricanes are giant whirlwinds, an idea that was soon supported by observations made by the British engineer William Reid, while surveying damage done by an 1831 hurricane in Barbados. Reid supplemented his account of the damage in Barbados with wind observations recorded in ships' logs, confirming the circular nature of the wind field.

Redfield and others believed that hurricanes extend upward only about a mile or so into the atmosphere. Their reasoning was based in part on the observation that hurricanes are often severely disrupted by relatively small mountains or hills. But in Cuba, Benito Viñes, a Jesuit priest and trained physicist, argued that hurricanes extend many miles upward, since they produce large amounts of cirrus. Such clouds are composed of ice crystals that can only form in the extremely cold conditions high up in the atmosphere. This controversy about the height of hurricanes was not finally settled until the 1930s, when the laws of physics were sufficiently well-known to rule out the idea of shallow hurricanes. Today we know that most hurricanes extend through the whole depth of the troposphere and into the lower stratosphere, sometimes reaching heights of around 18 km (11 mi).

Figure 2.1: Satellite image of Hurricane Floyd approaching the east coast of Florida in 1999. The image has been digitally enhanced to lend a three-dimensional perspective.

▲

The advent of reconnaissance aircraft and radar in the 1940s, and of satellites in the 1960s, gave scientists the means to determine the detailed structure and evolution of hurricanes. Today, almost everything we know about hurricanes is based on observations by aircraft, radar, and satellites.

Figure 2.1 shows a satellite picture of Hurricane Floyd approaching the east coast of Florida in 1999. The image has been enhanced using a computer, to give a feeling for the relative heights of clouds in the storm. The swirling mass of clouds is about 300 km (180 mi) across, and Floyd's eye is about 50 km (30 mi) in diameter. Surrounding the eye is a deep ring of thick cloud called the *eyewall*. A thin veil of very high cloud covers most of the swirling mass of thicker clouds; this is the veil of cirrus, made up of tiny

ice crystals, that Benito Viñes spoke of. Underneath it, and thus invisible from above, are several spiral-shaped bands of deep, thick cumulonimbus clouds, interspersed between relatively clear sky. These spiral bands produce heavy rains and gusty winds well outside the eyewall of the hurricane, where the strongest winds are found.

Let's have a closer look at the eye. Figure 2.2 shows a photo taken from a U.S. space shuttle overpass of Hurricane Emilia in the eastern North Pacific in 1994. Looking at the eye from above is like staring down the middle of a bathtub whirlpool, except that the boundaries are made of cloud and the "funnel" comes to an abrupt end at the sea surface. The eyewall is visible as the stadium of thick, white cloud surrounding the eye. This is where the strongest winds and heaviest rain are found. Unlike the

Figure 2.2: The eye of Hurricane Emilia over the eastern North Pacific, as seen from directly above by the crew of the space shuttle Columbia *on July 19, 1994, at 19:33 UT. At this time, Emilia had maximum winds of 70 m/s (155 mph).*

▶

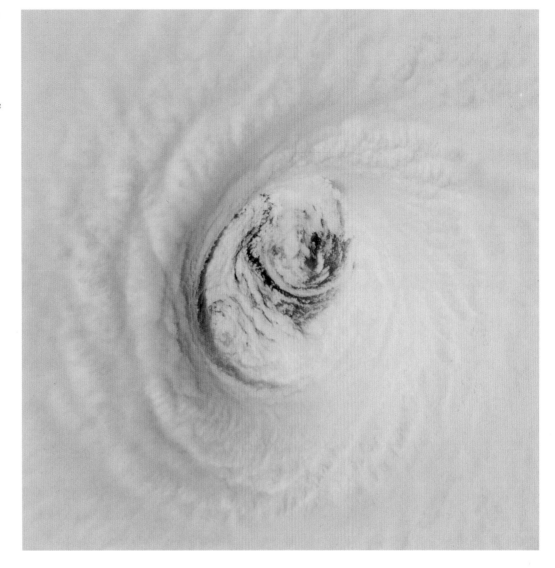

bathtub whirlpool, the air inside the eyewall is going up, not down, as it swirls around the eye. There are spiral striations along the inner edge of the eyewall and two big, swirling eddies in the low clouds at the base of the eye.

No mere photograph can do justice to the sensation of being inside the eye of a hurricane. Imagine a Roman coliseum 20 mi wide and 10 mi high, with a cascade of ice crystals falling along the coliseum's blinding white walls. The photo displayed in Figure 2.3, taken from a reconnaissance aircraft, gives some impression of the beauty of a hurricane's eye.

However impressive a hurricane's visual appearance, it is the instruments aboard satellites and aircraft that reveal the essence of a hurricane's structure and inner workings. In the early days of aircraft reconnaissance, the crew estimated wind speeds by looking down and noting how rough the sea appeared. Today, a sophisticated array of instruments automatically collects large quantities of high-quality data, which are taken back to laboratories and analyzed.

Among the most essential equipment is the meteorological radar, which works by transmitting pulses of electromagnetic radiation and measuring the radiation that is scattered back to the radar from raindrops, snowflakes, hailstones,

Figure 2.3: The eye of Hurricane Georges in photo taken from a reconnaissance aircraft. The eyewall at right is casting a shadow across part of the eyewall ahead.

▶

and other forms of precipitation. By measuring the time it takes the radiation to return, one can calculate the distance to the scattering particles, and by measuring how much radiation returns (and correcting for distance), one can establish the concentration of scatterers. Big particles, like hailstones, are much more efficient scatterers than smaller particles, like raindrops; and the tiny droplets and ice crystals that comprise what we see as cloud are too small to scatter a detectable amount of radiation. Thus what we see on a radar display is a measure of the concentration of precipitation in the air, weighted toward the bigger raindrops and ice particles. The radar antenna sweeps around in a circle, all the while transmitting and receiving pulses of radiation, and can thus in a few seconds survey a circular area several hundred kilometers in diameter.

The image displayed in Figure 2.4 was made using a radar mounted on an aircraft flying in Hurricane Floyd. It shows the distribution of precipitation-size scatterers about 1 km (3,300 ft)

above the surface. The heaviest rain is indicated by yellow and orange, with progressively lighter rain indicated by greens and blues. Floyd's eye is the circular blue region near the center of the image, and it is surrounded by the intense rain of the eyewall. In this case, the eyewall is not perfectly circular but consists of several arc-shaped regions of very heavy rain. Outside the eyewall is a region of reduced precipitation, known as the *moat;* this is surrounded by another ring of heavy rain that may constitute an *outer eyewall.* Part of this ring is composed of a spiral band that is wrapping inward from the south to the east of the storm center. There is also a prominent spiral band that starts north of the outer ring and extends eastward and then southward.

Figure 2.5 shows an east-west vertical slice through the storm made about the same time as the image in Figure 2.4. The figure spans a distance of 120 km (75 mi) centered at the storm center and extends upward to 20 km (12.5 mi) altitude. Note the pronounced eyewall of high

Figure 2.4: Radar reflectivity map from a hurricane recon-naissance aircraft flying in the eye of Hurricane Floyd of 1999. Map is 360 km (225 mi) square. The radar reflectivity measures roughly how much rain, snow, and hail are in the air.

▶

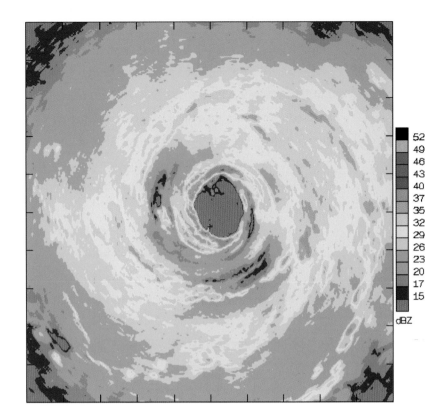

Figure 2.5: Vertical cross section of radar reflectivity in Hurricane Floyd of 1999, made from hurricane recon-naissance aircraft in the eye, located at the "+" sign. The diagram spans 20 km (12.5 mi) in height and is 120 km (75 mi) across.

▶

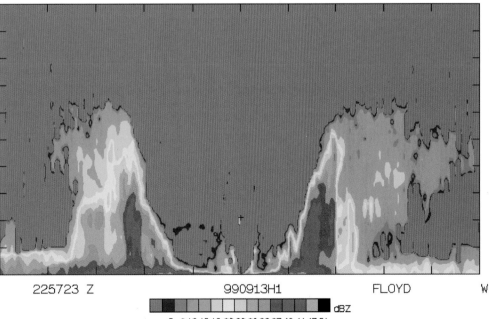

radar reflectivity, sloping outward with height. Since the vertical scale of this cross section is exaggerated, the real slope of the eyewall is nearly 1 to 1. Also note that the eye is not entirely free of radar returns. There are quite a few echoes near the sea surface; perhaps this is sea spray. Also, a peculiar, eyewall-like structure is present near the very center of the storm but extends only up to about 2 km (6,500 ft).

Outside the eyewall, radar echoes reach outward, centered at about 10 km (6 mi) altitude. These radar echoes represent *stratiform precipitation*—mostly snow and ice formed in horizontally layered clouds that flow out of the tall cumulonimbi of the eyewall. There are also radar returns very near the surface, revealing rain from shallower clouds and, perhaps, some sea spray.

In addition to carrying radar, aircraft are also used to directly measure meteorological quantities such as temperature, pressure, and humidity. Determining wind direction and speed requires a little more doing. Sensors on the aircraft measure its speed through the air, but to find the wind direction and speed with respect to the ground, one has to first determine the speed of the aircraft relative to the ground. Beginning in the late 1950s, this was done by pointing a radar at an angle to the ground. The radar measures the Doppler shift of the radiation reflected from the ground and thereby determines how fast the aircraft is moving with respect to the surface. (The same principle can be applied to sound. As a train goes by, the pitch of its whistle decreases; by measuring how much the pitch decreases, you can determine how fast the train is going.) The Doppler technique is not ideal: for one thing, it can be fooled by rain, sea spray, and ocean waves. Starting in the 1960s, aircraft motion was determined by a system of gyroscopes that very carefully keep track of all the accelerations experienced by the aircraft. By integrating these accelerations over time, one can find the aircraft's velocity over the ground, and by integrating these velocities in time, one can determine the aircraft's location. This system, called "inertial navigation," was used for some time to navigate aircraft. Today, the satellite-based Global Positioning System (GPS) is the method of choice for finding the aircraft's position and speed over the ground.

Figure 2.6 shows composite views of the structure of a hurricane revealed by direct measurements from research aircraft. These diagrams were made by fitting a satellite image of the top of one hurricane (Fran of 1999) to aircraft measurements of wind, temperature, and humidity from a different storm (Inez of 1966). The top panel shows the speed of the wind going counterclockwise around the center of the storm; the black dashed lines show the altitudes at which the aircraft were flying. (These lines are also displayed in the other two panels in this figure.) The darkest shade of blue indicates the lightest winds, less than 5 m/s (11 mph). Each successive color represents an increment of 5 m/s (11 mph), with the bright orange showing winds in excess of 65 m/s (145 mph). The strongest winds occur just above the ocean in a ring around the center, slightly outside the eyewall of the storm. Winds decrease rapidly inward toward the center, which is practically calm. They also decrease outward from the ring of maximum wind, but much more slowly, so that strong winds can still be found 100 km (60 mi) from the center of the storm. As we go up from the surface, the counterclockwise winds get lighter and lighter. If the highest aircraft had flown even higher, it would have actually encountered "negative" wind speed, which here means winds going the other way (clockwise) around

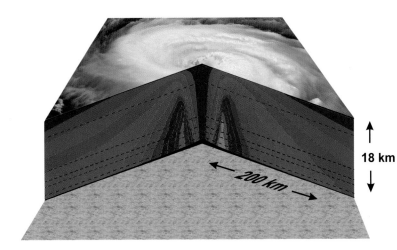

18 km

← 200 km →

Figure 2.6: Composite views of the distribution of various quantities in a hurricane, as revealed by measurements made from research aircraft. In each panel, the center of the hurricane is in the center of the panel, and the cutaway extends outward about 200 km (125 mi) and upward to about 18 km (11 mi).

Top panel: wind speed; middle panel: difference between temperature in the storm and temperature at the same altitude far away from the storm; bottom panel: entropy, a measure of the heat content of the air.

◀

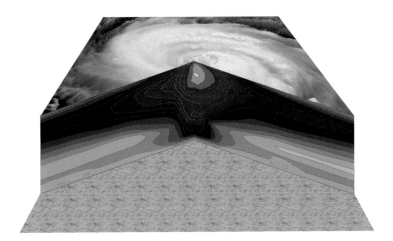

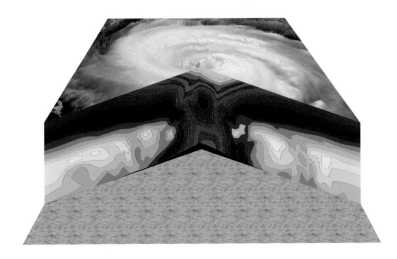

the center. Near the top of the storm, the clockwise winds increase from the center, reaching maximum speeds several hundred kilometers from the eye. But whereas the actual winds are nearly circular in the lower part of the storm, the high-level clockwise flow is usually very irregular, being concentrated in a few outward-curving jets.

The middle panel shows the difference between the temperature at a given point and the temperature at the same altitude far away from the storm. The lightest shade of blue shows no temperature difference at all, while bright yellow shows air that is more than 16°C (29°F) warmer than its distant environment; each shade represents an increment of 1°C (1.8°F). Hurricanes are very hot at their centers, especially near their tops. This great warmth is one consequence of the enormous amount of heat sucked out of the ocean by the storm. On the other hand, there is very little change in temperature as you move in toward the eye along the surface of the ocean.

The distributions of temperature and wind in hurricanes are related to each other. Look carefully at the upper two panels of the figure. Notice that the rate at which the wind decreases upward is proportional to the rate at which the temperature decreases outward. This relationship is fundamental to rotating fluid flows and is called the *thermal wind relation.*

The lowest panel in the figure shows a quantity called entropy, which is a measure of how much heat has been added to the air. The smallest values of entropy are denoted by the light blue colors, while yellow represents the largest values. There are two main sources of entropy in a hurricane: heat transfer from the ocean to the atmosphere, which is accomplished mostly by evaporation, and frictional dissipation of wind energy. Heat transfer by evaporation is familiar to anyone who has ever stepped out of a swimming pool on a windy day. Even when the air is hot, you feel chilled by the wind. This is because the evaporation of water from your skin takes heat from your body. This heat does not disappear: it is added to the air in the form of *latent heat,* which is proportional to the amount of water vapor in the air. When this water vapor later condenses into cloud, the latent heat is released to the air, warming it in the process. Frictional heating is sometimes used to start campfires. When two sticks are rubbed vigorously together, some of the energy of the motion of the sticks is converted into heat energy. Similarly, the friction of air rubbing the ocean surface, or of quantities of air or water sliding past one another, converts some of the energy of motion into heat. Both the transfer of heat from the ocean and the frictional heating increase the entropy of air flowing inward toward the hurricane's eyewall in the lowest kilometer of the atmosphere. Then, as air enters the eyewall, it turns upward and slightly outward, spiraling upward through the eyewall and carrying with it the large entropy it acquired on its way into the eyewall. Near the top of the storm, the high-entropy air turns outward. Ultimately, the heat acquired from the ocean in a hurricane is exported to the distant environment, where it is finally lost by radiation to space.

Mathematical models provide another potential source of information about hurricanes. These models use computers to solve the differential equations that govern the evolution of the atmosphere. These equations are based on Newton's law applied to fluids, the first law of thermodynamics, and the conservation of water in its various phases. Unfortunately, computers are not yet powerful enough to accurately calculate certain physical processes in hurricanes, including turbulence, the formation and fall of rain, and the intricate interaction of radiation with clouds. These processes must be represented by approximations. Moreover, even today's computers can't keep track of every molecule of air and must work instead using fairly sizable chunks of air. In even the most advanced models, these chunks of air are many cubic kilometers in volume. The inaccuracies introduced by dividing up the air in this way, and by the approximations to the physics, render the models imperfect. In the end, one can only have confidence in the models to the extent that their results compare favorably to observations of hurricanes.

Yet models are good at filling in details that are difficult to observe. For example, while aircraft can make excellent measurements of the horizontal motion of air, measuring the air's vertical motion (updrafts and downdrafts) proves far more difficult. Updrafts and downdrafts tend to be widely separated, and an airplane may only sample a few on its way across a storm. The average vertical air motion calculated from such measurements is unlikely to correspond to the true

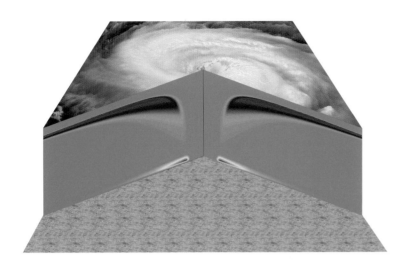

Figure 2.7: The same cross sections as in Figure 2.6, except showing quantities generated by a computer model. Top panel: radial air motion, with yellow and orange denoting inflow and blue and violet denoting outflow. Bottom panel: upward air motion.

◄

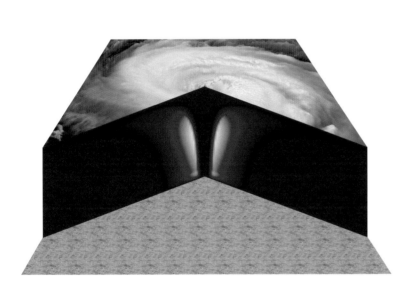

results of calculations using a mathematical model. It shows inward flow next to the sea surface (yellow and orange), strong upward motion in the eyewall, and outflow in a thin layer near the top of the storm (blue and violet). Although it is too weak to be seen in this figure, there is downward motion in the eye itself, strongest right next to the eyewall, and also in a broad region outside the eyewall. But since this figure is made by averaging the air motion along circles centered at the storm center, the strong updrafts associated with spiral rain bands are missed.

The nature of the airflow and the distribution of clouds in a typical hurricane are summarized in Figure 2.8. Air spirals in toward the eyewall in the lowest kilometer or so, counterclockwise in the Northern Hemisphere and clockwise in the Southern Hemisphere. The spiral becomes tighter and tighter near the eyewall, as the wind speed increases to hurricane force. Some of the inward-spiraling air is drawn into the strong updrafts associated with spiral rain bands outside the core of the storm, but most of it converges into the eyewall, where it turns abruptly upward and ascends in a broadening spiral to the top of the storm. There the airflow turns progressively outward, and after spiraling out a few hundred kilometers, its rotation

average. Here the computer model has the advantage that it can make exact "measurements" of vertical motion. So we have to choose between imperfect measurements of the real thing and perfect measurements of the fiction generated by a mathematical model.[1] The best we can do is to draw conclusions from a judicious blend of the two.

Figure 2.7 shows composite views of the vertical and radial (in-out) air motion in a hurricane, similar to Figure 2.6 but now showing the

---

[1] Or, as one wit put it, we must choose between observing the unpredictable and predicting the unobservable.

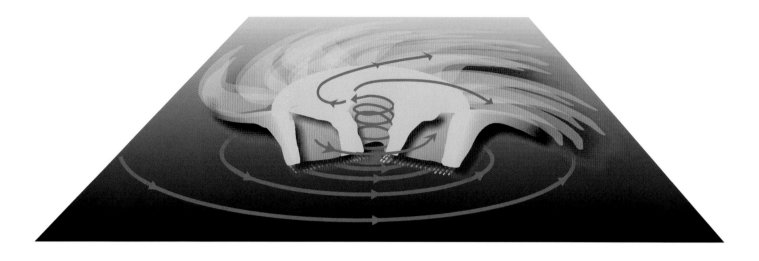

Figure 2.8: Synopsis of the airflow and cloud distribution in a mature Northern Hemisphere hurricane.

▲

reverses, turning clockwise in the Northern Hemisphere. But unlike the spiraling inflow, air coming out of the top of a hurricane seldom retains much circular symmetry. Instead, it tends to concentrate in one or more "outflow jets" that carry away the exhaust of this massive heat engine. The entire system moves as one body across the surface of the ocean, and the speed of its movement is added to winds on one side of the storm, and subtracted from them on the other side.

The hurricane, once fully developed, is among the most coherent and persistent structures that inhabit the otherwise chaotic atmosphere of our planet. However terrible its effects, one cannot help but admire the intricate beauty of its architecture.

▲

The sailing vessel *Ouragan* in a hurricane. Engraving from *Les Meteores*, Paris, 1869.

I have seen tempests, when the scolding winds
Have riv'd the knotty oaks; and I have seen
The ambitious ocean swell and rage and foam,
To be exalted with the threat'ning clouds:
But never till to-night, never till now,
Did I go through a tempest dropping fire.
Either there is a civil strife in heaven,
Or else the world, too saucy with the gods,
Incenses them to send destruction.

—William Shakespeare (1564–1616),
from *Julius Caesar*, Act I, Scene iii

3   Huracán

Hurricane: A name given primarily to the violent wind-storms of the West Indies, which are cyclones of diameter of from 50 to 1000 miles, wherein the air moves with a velocity of from 80 to 130 miles an hour round a central calm space, which with the whole system advances in a straight or curved track; hence, any storm or tempest in which the wind blows with terrific violence.

—Oxford English Dictionary

Hurricane: (Many regional names.) A tropical cyclone with 1-min average surface (10 m) winds in excess of 32 m s⁻¹ (64 knots) in the Western Hemisphere (North Atlantic Ocean, Caribbean Sea, Gulf of Mexico, and in the eastern and central North Pacific east of the date line). The name is derived from "huracán," a Taino and Carib god, or "hunraken," the Mayan storm god.

—American Meteorological Society Glossary of Meteorology

The word *hurricane* comes to us via the early Spanish explorers of the New World, who were told of an evil god of winds and destruction, variously called Huracán, Hunraken, or Jurakan in the Caribbean and Mexico. In the legends of the Mayan civilizations of Central America and the Tainos of the Caribbean, this god played an important role in their Creation. According to Taino legend, the goddess Atabei first created the earth, the sky, and all the celestial bodies. To continue her work, she bore two sons, Yucaju and Guacar. Yucaju created the sun and moon to give light, and then made plants and animals to populate the earth. Seeing the beautiful fruits of Yucaju's work, Guacar became jealous and began to tear up the earth with a powerful wind, renaming himself Jurakan, the god of destruction. Yucaju then created Locuo, a being intermediate between a god and a man, to live in peaceful harmony with the world. Locuo, in turn, created the first man and woman, Guaguyona and Yaya. All three continued to suffer the powerful winds and floods inflicted by the evil Jurakan.

The natives of the Caribbean were clearly terrified of Jurakan, and would shout, beat drums, and engage in bizarre rituals in attempts to drive the god away. The early inhabitants of Cuba carved images in stone of the god they called Huracán; two such images are reproduced in Figures 3.1 and 3.3.

The various likenesses of Huracán invariably consist of a head of indeterminate gender with no torso, and two distinctive arms spiraling out from its sides. Most of these images exhibit cyclonic (counterclockwise) spirals. The Cuban ethnologist Fernando Ortiz believes that they were inspired by the tropical hurricanes that have always plagued the Caribbean. If so, the Tainos discovered the vortical nature of hurricanes many hundreds of years before the descendents of European settlers did. How they may have made this deduction remains mysterious. The spiral rain bands so well known to us from satellite pictures were not "discovered" until meteorological radar was developed during

Figure 3.1: Likeness of the god Huracán, from a Cuban ceramic vase.

Figure 3.2: Universal symbol of the tropical cyclone. In the Southern Hemisphere, the arms are curved in the opposite sense.

▶

Figure 3.1: Likeness of the god Huracán, from a Cuban ceramic vase.

Figure 3.2: Universal symbol of the tropical cyclone. In the Southern Hemisphere, the arms are curved in the opposite sense.

World War II, and they are far too big to be discerned by eye from the ground. Perhaps these ancient people surveyed the damage done by Huracán and, based on the direction trees fell, concluded that the damage could only have been done by a rotating wind, as did William Redfield in 1831. Or perhaps they witnessed tornadoes or waterspouts, which are much smaller phenomena whose rotation is readily apparent, and came to believe that all destructive winds are rotary.

Whatever its source, the image of Huracán remains an enigmatic premonition of what would be seen in radar and satellite imagery many hundreds of years later, and what would form the basis of the internationally recognized symbol of a tropical cyclone, shown in Figure 3.2.

The word *huracán* evolved through several variations (Shakespeare's King Lear cries, "Blow, winds, and crack your cheeks! rage! blow! You cataracts and hurricanoes, spout Till you have drench'd our steeples, drown'd the cocks!"). Today *hurricane* has a very specific meaning. It is one of many regional names given to an intense form of a general phenomenon known as a *tropical cyclone,* a low-pressure area that forms over tropical oceans and is associated with cyclonically rotating winds through most of the atmosphere. (Cyclonic rotation is counterclockwise in the Northern Hemisphere and clockwise in the Southern Hemisphere.) The table on page 21 shows a more precise definition of the term, together with subclassifications and regional names.

For a tropical cyclone to technically qualify as a hurricane, it must have winds of at least 33 m/s (74 mph) and occur over the Atlantic, eastern North Pacific, or eastern South Pacific ocean. But in this book, I shall sacrifice technical accuracy for the sake of brevity by using the popular term *hurricane* rather than *tropical cyclone* to denote the generic phenomenon, though where the distinction is crucial, I shall resort to the latter.

The word *cyclone* itself was coined in 1848 by the Englishman Henry Piddington, then curator of the Calcutta Museum in India and later president of the Marine Courts of Inquiry. Piddington, who was keenly interested in storms affecting India, derived the term from a Greek word meaning "coil of a snake."

The precise origin of the word *typhoon* is, however, controversial. It very likely originated from the Chinese *jufeng*. *Ju* can mean either "a wind coming from four directions" or "scary"; *feng* is the generic word for wind. Arguably the first scientific description of a tropical cyclone and the first appearance of the word *jufeng* in the literature is contained in a Chinese book entitled *Nan Yue Zhi* (Book of the Southern Yue Region), written around A.D. 470. In that book, it is stated that "Many

Figure 3.3: Huracán, *by Rafael Tufiño. The shape in the lower central part of the print is a* cemí, *a figure carved in stone, believed to represent benign spirits but perhaps also symbolizing the island of Puerto Rico, which is mountainous in its interior.* Jurakan *floats ominously over this* cemí, *blowing into his conch shell to produce hurricane winds. The decorative border at the bottom is made of frog figures, commonly used by the Taino to represent flooding, while the curved streaks throughout may represent strong winds. Other symbols represent the* coquí *and the sun, among other things.*

▶

jufeng occur around Xi'an County. *Ju* is a wind (or storm) that comes in all four directions. Another meaning for *jufeng* is that it is a scary wind. It frequently occurs in the sixth and seventh month.[2] Before it comes, roosters and dogs are silent for three days. Major ones may last up to seven days. Minor ones last one or two days. These are called *heifeng* (black storms/winds) in foreign countries."[3]

The Cantonese *tai-fung* and the Mandarin *ta-feng* derive from *jufeng*. European travelers to China in the sixteenth century noted that a word sounding like *typhoon* was used to denote severe coastal windstorms. On the other hand, *typhoon* was used in European texts around 1500, long before systematic contact with China was established. It is possible that the European use of this word was derived from *Typhon*, the draconian earth demon of Greek legend. Son of Gaia and Zeus, this god was said to be of human form above the waist, but with legs of hissing vipers. Later Greeks referred to Typhon as a god of evil winds, and the Arabic word *tufan* denotes a violent wind and is probably derived from *tafa*, to turn around.

[2] Of the Chinese lunar calendar; roughly July and August of the Gregorian calendar.

[3] I am grateful to Professor Kam-Bui Liu of Louisiana State University for pointing this out. Professor Liu is one of the world's authorities on the history of typhoons striking China.

By the early eighteenth century, *typhon* and *typhoon* were in common use in European literature, as in the poem "Summer" by the Scottish poet James Thomson (1700–1748):

> *Beneath the radiant line that girts the globe,*
> *The circling Typhon, whirled from point to point.*
> *Exhausting all the rage of all the sky,*
> *And dire Ecnephia, reign.*

TERM	DEFINITION	REGION USED
Tropical Cyclone	Nonfrontal synoptic-scale (200–2,000 km in diameter) low-pressure system originating over tropical or subtropical waters with organized convection (i.e., rain shower or thunderstorm activity) and definite cyclonic surface wind circulation	Global
Tropical Depression	A tropical cyclone with *maximum sustained surface winds*[4] of less than 17 m/s (39 mph)	Global
Tropical Storm	A tropical cyclone with *maximum sustained surface winds*[4] of at least 17 m/s (39 mph) but less than 33 m/s (74 mph)	Global
Hurricane	A tropical cyclone with *maximum sustained surface winds*[4] of at least 33 m/s (74 mph)	North Atlantic Ocean, Northeast Pacific Ocean east of the International Dateline, and the South Pacific Ocean east of 160°E
Typhoon	"	Northwest Pacific Ocean west of the International Dateline
Severe Tropical Cyclone	"	Southwest Pacific Ocean west of 160°E and Southeast Indian Ocean east of 90°E
Severe Cyclonic Storm	"	North Indian Ocean

[4] Unfortunately, there are at least two definitions of this term in widespread use. Most countries use the World Meteorological Organization's definition, which is a ten-minute average wind at an elevation of ten meters. But the United States uses a one-minute average.

▲

Hurricane, Bahamas.
Watercolor by American
artist Winslow Homer,
1888–89.

I think that the Root of the Wind is Water—
It would not sound so deep
Were it a Firmamental Product—
Airs no Oceans keep—
Mediterranean intonations—
To a Current's Ear—
There is a martime conviction
in the Atmosphere—

     —Emily Dickinson (1830–1886)

4   The Tropical Hothouse

Warmest climes but nurse the cruellest fangs.

—Herman Melville, *Moby Dick*

"

n all the Indies, I have always found May-like weather," said Christopher Columbus of his first voyage to the New World. A quick glance at the travel section of any newspaper reveals the popularity of tropical resorts. We love the Tropics for their dependable, salubrious climates: warm, steady breezes; plentiful sunshine; and rain showers that seldom last more than a few minutes. In places like Tahiti, the minimum temperature in winter is only a few degrees colder than the maximum temperature in summer, and the nights are almost as warm as the days.

Why, then, do such peaceful, benign climates engender the most terrible storms on earth?

The answer, in short, is the greenhouse effect. By trapping heat energy in the ocean, the greenhouse effect sets the stage for the meteorological explosion that is the hurricane.

Almost all the energy that heats the earth and powers the winds comes from the sun. Sun-light is a form of radiation, consisting of waves that travel through space as well as through various kinds of matter, such as air. As Figure 4.1 shows, sunlight contains many different wavelengths, many of which are contained within the narrow range we see as visible light. White light is actually the combination of light with a range of colors, each of which corresponds to a different wavelength. Red light, for example, has a wavelength of about 0.7 microns. (A micron is one-millionth of a meter; there are 25,400 microns in one inch.) Blue light has a shorter wavelength, of about .45 microns.

But as Figure 4.1 shows, only a small fraction of the whole spectrum of radiation is visible. Wavelengths that are too short to see, beyond the blue end of the visible spectrum, comprise ultraviolet radiation (the form of radiation that causes sunburn). Gamma rays and x-rays are forms of radiation with even shorter wavelengths. Wavelengths too large to see (0.7–500 microns)

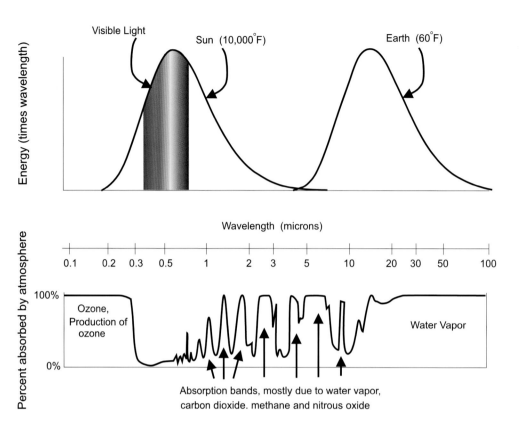

Figure 4.1: The top panel shows the amount of energy per unit wavelength (here multiplied by wavelength) as a function of the wavelength itself, which is given by the axis in the middle. The wavelengths are given in microns: one micron is one millionth of a meter. The curve at the left shows the energy content of sunlight; the colors denote the visible part of the spectrum. The curve at the right shows the spectrum of terrestrial, or infrared, radiation. The bottom panel shows the percentage of energy that is absorbed at each wavelength by various gases in the atmosphere.

►

comprise infrared radiation. Radio and television transmissions are carried by even longer wavelengths.

All matter that has a temperature above absolute zero emits radiation. For example, the human body emits radiation concentrated at wavelengths between about 20 and 50 microns. The hotter the matter, the shorter the wavelength of radiation it emits. For example, the sun, with a surface temperature of about 6,000°C (10,000°F), emits radiation primarily in the 0.2–3.0 micron range, whereas the earth emits infrared radiation, with wavelengths primarily between 5 and 50 microns.

Since the sun is so much hotter than the earth, there is almost no overlap between the radiation emitted by the two bodies (see Figure 4.1). For this reason, scientists who study the

atmosphere call solar radiation *shortwave radiation*, while they refer to the earth's radiation as *longwave radiation*.

The total amount of radiant energy an object emits depends on its physical properties, most important, its temperature. This is expressed by the *Stefan-Boltzmann equation*,

$$\varepsilon\sigma T^4,$$

in which σ is a universal constant (called the Boltzmann constant), T is the absolute temperature of the emitting body, and ε is the emissivity of the body, which may vary between zero and one. (An object that emits radiation perfectly is called a blackbody and has an emissivity of one.) So the amount of energy an object radiates rises quickly with its temperature.

If the earth had no atmosphere, we could easily calculate its temperature. We would first

Figure 4.2: Cartoon of a simple molecule consisting of two atoms. The primary gases of our atmosphere, nitrogen (N$_2$) and oxygen (O$_2$), have this form.

▶

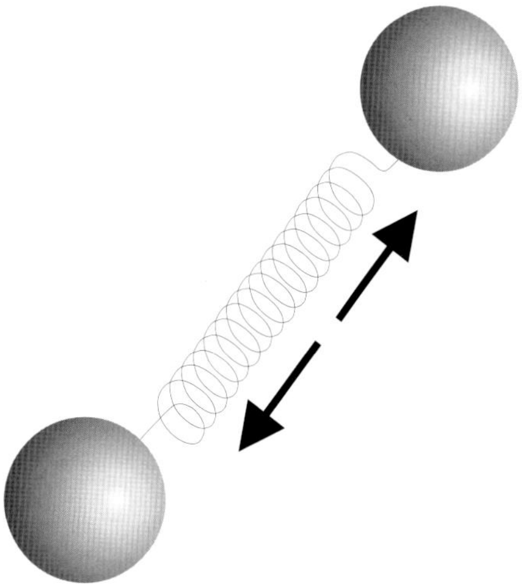

calculate the amount of energy, E, received from the sun as shortwave radiation, allowing for the reflection of some of it back to space. Then we would use the Stefan-Boltzmann equation (p. 24) to determine how hot the earth would have to be to radiate the absorbed solar energy back to space. (If it radiated less, the earth would warm with time; if it radiated more, it would cool.) Doing the calculation gives a temperature of only about 0°F.

Yet the average temperature of the earth's surface is close to 60°F. This relative warmth is owing to the greenhouse effect of the earth's atmosphere. How does this greenhouse effect work?

About 98 percent of our atmosphere is composed of molecules of nitrogen and oxygen, each having two atoms. They can be thought of as two spheres connected by a spring, as shown in Figure 4.2.

These are simple molecules. About all they can do is stretch and compress. To absorb or emit radiation, a gas molecule has to be able to change its own energy content, for example, by changing the rate at which it stretches and compresses. Since oxygen and nitrogen molecules can move in only a limited way, they cannot absorb or emit much radiation.

Fortunately, our atmosphere contains more versatile molecules. Among the most important is water vapor, or H_2O (one oxygen atom and two hydrogen atoms), whose form is illustrated in Figure 4.3.

Compared to molecular nitrogen and oxygen, water vapor molecules are capable of great gymnastic feats. Besides being able to stretch and compress, they can bend at their midsections, rotate, and perform combinations of stretching, bending, and rotating. Because they can move in such complex ways, they can absorb and emit much more radiation than molecules that consist of only two atoms.

Each time a molecule absorbs radiation, its energy state increases, and each time it emits radiation, its energy state decreases. But changes in the energy state of a single molecule are

Figure 4.3: Cartoon of a tri-atomic molecule, such as water vapor (H_2O) or carbon dioxide (CO_2).

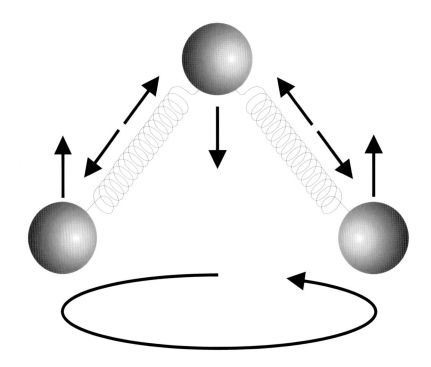

quickly communicated to neighboring molecules with which it collides. Thus, absorption of radiation ultimately makes the molecules travel faster in between collisions with their neighbors. Temperature is proportional to this rate of travel. Absorption of radiation, then, increases air temperature, while emission of radiation decreases it.

In addition to water vapor, several other molecules consisting of three or more atoms are important for absorbing and emitting radiation in the atmosphere: carbon dioxide (CO_2), ozone (O_3), methane (CH_4), and nitrous oxide (N_2O) are the most important. Together, they comprise only about 1 percent of the mass of the atmosphere, yet they are important for setting the temperature of the earth and its atmosphere. Condensed water, in the form of clouds of water droplets or ice crystals, is also important in absorbing, emitting, and reflecting radiation.

Yet even these gases can only interact with certain discrete wavelengths of radiation. The amount of radiation absorbed at each wavelength is shown in the bottom panel of Figure 4.1.

Note that almost all of the solar part of the spectrum of radiation penetrates to the surface: *the atmosphere is almost transparent to solar radiation.* Only the ultraviolet part of the spectrum is absorbed, mostly by ozone. (Were there no ozone, lethal amounts of ultraviolet radiation would reach the surface.)

In contrast, *the atmosphere is nearly opaque to infrared radiation.* The radiation absorbed by gases in the atmosphere is re-emitted, both upward and downward. In consequence, the earth's surface receives both infrared radiation emitted downward from the atmosphere and solar radiation. It is a remarkable fact that, on average, the surface receives more radiation from the atmosphere than directly from the sun.

To balance this extra source of radiation, the earth's surface has to warm up so it can radiate more heat (by the Stefan-Boltzmann equation). A detailed calculation of the radiative balance of the earth's surface and atmosphere yields an average surface temperature of 135°F, much warmer than observed.

The problem with the greenhouse calculation described earlier is that radiation is assumed to carry *all* the heat energy through the atmosphere. In reality, air currents carry much of the heat, especially near the earth's surface. These air currents are called *convection*.

Ordinary convection carries warm air upward and cool air downward. Although such currents are invisible, they are responsible for many of the "bumps" experienced by passengers of aircraft landing or taking off on warm, sunny days.

By carrying heat upward, convection cools the earth's surface well below the temperature it would have if all the heat were carried by radiation. But there is another important reason why the earth's surface is cool: it is wet. Just as you feel cold when emerging from a swimming pool, even on a warm day, the earth's surface is cool because water is evaporating from it. It takes energy to evaporate water, and some of the radiation absorbed by the earth's surface is used to evaporate water rather than to heat the surface and the air just above it.

The heat energy used to evaporate water from the surface does not disappear. It remains in the forms of "latent" heat, which is proportional to how much water vapor is in the air. When the water vapor condenses into water droplets or ice crystals, this latent heat is released into the atmosphere, thereby warming it. In the Tropics, and over middle-latitude continents during the warmer months, this condensation occurs within upward convection currents, forming cumulus

and cumulonimbus clouds ("thunderheads"). The latent heat released when water vapor condenses in these clouds makes them warm, and their warmth drives them further upward.

The condensation of water vapor thus both powers the convection currents and makes them visible to us in the form of cumulus and cumulonimbus clouds. These convection currents transport both heat and water through much of the atmosphere. Taking into account the heat transported by convection as well as by radiation gives very reasonable estimates of the average surface temperature in the Tropics.

Figure 4.4 summarizes the processes that transport heat through the tropical atmosphere. Sunlight travels down through the atmosphere and is absorbed by the ocean (blue curve at right). If there were no convection currents, infrared radiation would transport heat upward at the same rate that sunlight transports heat down to the surface. But evaporation keeps the surface so cool that relatively little radiation is emitted by it; instead, convection (mostly in the form of cumulus clouds) takes most of the heat away from the surface, moving it upward and replacing it with cool, dry air from aloft (green curve). The higher one goes in the atmosphere, the less heat is transported by convection and the more is carried by infrared radiation, so that at and above about ten miles altitude, infrared radiation (red curve) is responsible for all of the upward heat transport.

In the normal state of the Tropics, the absorption of sunlight by the ocean is balanced not as much by the emission of infrared radiation as by the evaporation of water. But for ocean water to evaporate, the relative humidity of the air just above the sea surface must be less than 100 percent. (If it were 100 percent, there would be

Figure 4.4: The heat balance of the tropical atmosphere. The diagram at top illustrates the role of convection in transporting heat through the atmosphere. Warm, moist air ascends within tall cumulonimbus clouds, where the water vapor condenses into liquid water and ice. Some of the liquid water falls to the surface as rain. Cool, dry air descends between clouds. These currents serve to carry heat away from the earth's surface and deposit it in the upper atmosphere, where it is carried into space by infrared radiation. The diagram at bottom shows the way heat is transported by various processes. The blue arrow shows the downward solar radiation, the red arrow shows the heat carried upward by infrared radiation, and the green arrow shows the heat transported upward by convection.

▶

no evaporation because the air is already saturated and cannot "hold" more water.) In fact, the relative humidity is about 80 percent. The ability of ocean water to evaporate into the air is essential for supplying energy to a hurricane, as we shall see. Here we have seen that this evaporation is a direct consequence of the greenhouse effect: the more infrared radiation emitted downward by the atmosphere and clouds within it, the more evaporation is needed to get rid of the extra heat.

The greenhouse effect thus makes possible the greatest storms on earth.

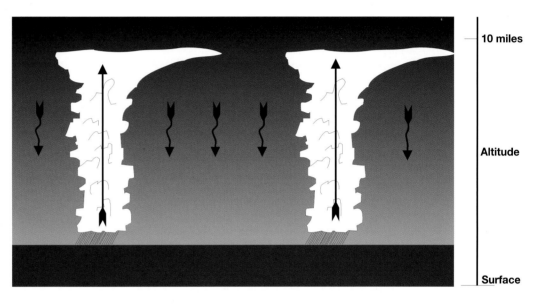

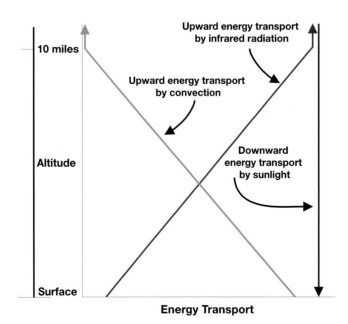

▲

Disaster to the Coast Survey Brig Peter G. Washington *in the Hurricane of September 8, 1846. Engraving from* Harper's New Monthly Magazine.

T hen seemed it that a tameless hurricane

Arose, and bore me in its dark career

Beyond the sun, beyond the stars that wane

On the verge of formless pace—it languished there,

And, dying, left a silence lone and drear,

More horrible than famine. In the deep

The shape of an old man did then appear,

Stately and beautiful; that dreadful sleep

His heavenly smiles dispersed, and I could wake and weep.

—Percy Bysshe Shelley (1792–1822), from Canto Third of
"The Revolt of Islam: A Poem in Twelve Cantos"

5   Columbus's Hurricane

From the strong Will, and the Endeavor
   That forever
Wrestle with the tides of Fate;
From the wreck of Hopes far-scattered,
   Tempest-shattered,
Floating waste and desolate;—

Ever drifting, drifting, drifting
   On the shifting
Currents of the restless heart;
Till at length in books recorded,
   They, like hoarded
Household words, no more depart.

—Henry Wadsworth Longfellow,
from "Seaweed"

Based on his first few voyages, Christopher Columbus concluded that the weather in the New World is benign: "In all the Indies, I have always found May-like weather," he commented. Although sailing through hurricane-prone waters during the most dangerous summer months, he did not have any serious hurricane encounters during his first three voyages. (But in June 1495, on his second voyage, three of his four ships were sunk in Isabela Harbor, on the south side of Hispaniola, by what seems to have been a waterspout.) During these voyages, Columbus heard from native inhabitants about horrible tempests they called "Huracán," the word for their god of evil (Chapter 3), and he probably learned from them the portents to look for.

By the time of his fourth voyage to the New World, in 1502, Columbus had had a serious falling-out with the bureaucrats appointed by Spain to govern the fledgling colonies and to extract gold and other commodities from the native inhabitants. Among the more unfriendly of these exploiters was Don Nicolas de Orvando, the governor of Hispaniola, with whom Columbus had been forbidden contact by his Spanish sovereigns. But as Columbus entered the outer roadstead of Santa Domingo, he recognized the ominous signs of an approaching violent storm: an oily swell emanating from the southeast and a developing veil of cirrostratus overhead. Concerned for the safety of his men and ships, he sent a message to Governor Orvando begging to be allowed to seek refuge in Santa Domingo harbor. Columbus had observed that the governor was preparing a large fleet of some 30 boats to set sail for Spain, carrying large quantities of gold and slaves, and warned him to postpone the trip and to take measures to secure the ships. Refusing both the request and the advice, Orvando read Columbus's note aloud to his minions, who roared with laughter at the forecast by the amateur meteorologist from Genoa. The laughter was short-lived. Orvando's ships left port, and as they

Figure 5.1: From Theodor De Bry's Reisen in Occidentalischen Indien *(Travels in the West Indies), travel accounts describing the Americas and some voyages around the world. This engraving was designed to accompany the text from Benzoni's* History of the New World, *which I have quoted on p. 32. De Bry, 1528–98, was a Dutch Protestant who became interested in travel accounts when he learned of the availability of the paintings of John White and Jacques le Moyne from the American colonies.* ▲

entered the strait between Hispaniola and Puerto Rico, the hurricane struck, sinking 25 ships on the spot and sending 4 more limping back to port, where they too sunk. Only one ship, the *Aguja*, made it to Spain, and that one, no doubt to Orvando's intense distress, was carrying what little remained of Columbus's own gold. Meanwhile, the weather-savvy admiral, anticipating strong winds from the north, positioned his fleet in a harbor on the south side of Hispaniola. On the thirtieth of June, the storm hit with ferocious northeast winds. Even with the protection of the mountainous terrain to windward, the fleet struggled. In the admiral's words, "The storm was terrible and on that night the ships were parted from me. Each one of them was reduced to an extremity, expecting nothing save death; each one of them was certain the others were lost." The anchors held only on Columbus's ship; the others were dragged out to sea, where their crews fought for their lives. Nevertheless, the fleet survived with minimal damage. Almost 18 months later, Columbus returned to Santo Domingo, only to discover that it had been largely destroyed by the hurricane.

The following describes an early post-Columbian encounter with a hurricane:

In those days a wondrous and terrible disaster occurred in this country. At sunrise such a horrible, strong wind began that the inhabitants of the island thought they had never seen or heard anything like it before. The raging storm wind (which the Spaniards call Furacanum) came with great violence, as if it wanted to split heaven and earth apart from one another, and hurl everything to the ground. All the people were so shocked by the storm of such unheard-of violence that they believed with fear and horror that death was wholly before their eyes, that the elements would completely melt [them], and [that] the last day was surely at hand. Just then it began to thunder and lightning frightfully, and it thundered so cruelly with cracks and crashes, and the lightning flashes came so quickly after one another that the sky seemed to be completely full of fire. Soon after that a thick and dreadful darkness came to the day which was even darker than any night could ever be, and no person could see the others for the darkness, but rather had to grope and fumble like the blind to find their way. The people were as a whole so despairing because of their great fear that they ran here and there, as if they were senseless and mad, and did not know what they did. Meanwhile the wind blew with such great and terrible force that it ripped many large trees out of the earth by the roots and threw them over. Similarly some large cliffs also fell down from the force with terrible, awful crashes and turmoil, so that many houses and villages were thrown to the ground including many people who stayed in that place. The strong and frightful wind threw some entire houses and capitals including the people from the capital, tore them apart in the air and threw them down to the ground in pieces. This awful weather did such noticeable damage in such a short time that not three ships stood secure in the sea harbor or came through the storm undamaged. For the anchors, even if they were yet strong, were broken apart through the strong force of the wind and all the masts, despite their being new, were crumpled. The ships were blown around by the wind, so that all the people in them were drowned. For the most part the Indians had crawled away and hidden themselves in holes in order to escape such disaster.

—Girolamo Benzoni, *History of the New World,* 1565

Tropical clouds near the
Florida Keys. Photograph
by Howard Bluestein.

I bring fresh showers for the thirsting flowers,
        From the seas and the streams;
I bear light shade for the leaves when laid
        In their noon-day dreams.
From my wings are shaken the dews that waken
        The sweet buds every one,
When rocked to rest on their mother's breast,
        As she dances about the sun.
I wield the flail of the lashing hail,
        And whiten the green plains under,
And then again I dissolve it in rain,
        And laugh as I pass in thunder.

        —Percy Bysshe Shelley (1792–1822), from "The Cloud"

6   The Tropical Downpour

Now the storm begins to lower,
(Haste, the loom of Hell prepare.)
Iron-sleet of arrowy shower
Hurtles in the darken'd air.

—Thomas Gray, from "The Fatal Sisters: An Ode"

In Chapter 4, I described why cumulus convection occurs in the Tropics and how it serves to transport heat and moisture away from the ocean surface. Now let's have a closer look at the individual clouds that do this work.

Everyone who lives in maritime tropical climates, or who as had a holiday in the Tropics, is familiar with the "tropical downpour." Nice weather lasts for many hours or days, but eventually, an impressive cumulonimbus cloud, often capped with a cirrus "anvil," bears down, usually from the east. An example of such a cumulonimbus cloud is shown on p. 33. As the cloud approaches, the sky near the horizon darkens, the wind shifts and blows away from the storm, and the air temperature drops by several degrees, providing temporary relief from the heat and humidity. Then the rain comes in torrents, usually lasting only a few minutes, and it is often followed by an hour or so of light, steady rain. Then it is all over. On land, these tropical showers are most frequent in the late afternoon, but over the ocean, they occur more frequently at night or early morning.

In the late 1940s, researchers fat with surplus aircraft from World War II conducted a series of investigations of ordinary convective showers and thunderstorms, by flying through the clouds at various stages of their development. The "Thunderstorm Project," as it was called, described in detail the structure and evolution of convective showers, characterizing their life cycle as evolving through three stages, as shown in Figure 6.1.

In the first, or "cumulus" stage, a draft of particularly warm, moist air from near the surface rises through the slightly cooler air around it. (Warm air is less dense than cold air, and moist air is also less dense than dry air at the same temperature. A pocket of less dense air is accelerated upward through the denser air surrounding it, a principle discovered by the Greek natural

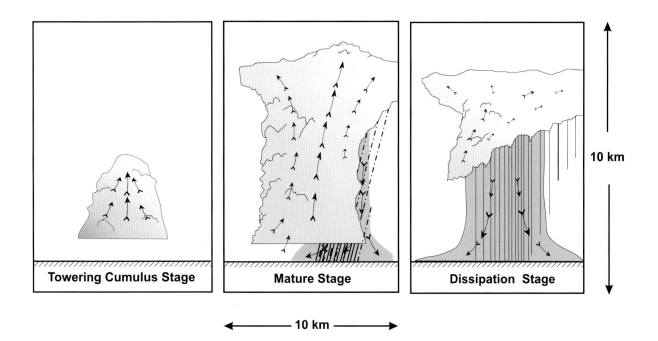

Towering Cumulus Stage   **Mature Stage**   **Dissipation Stage**

10 km

10 km

Figure 6.1: Three stages in the development of a typical tropical rain shower. During the first stage (left), a cumulus cloud grows upward, but precipitation has not yet formed. After 20 minutes or so, rain begins to fall as the mature stage is reached (middle). Some of the rain evaporates, chilling the air and causing a downdraft that spreads out at the surface. After another half hour or so, this spreading, rain-cooled air cuts off the supply of warm, moist air feeding the storm, which then begins to dissipate (right).

▲

philosopher Archimedes more than two thousand years ago and applied by hot air balloonists.) As the air rises, it experiences lower and lower pressure. In response to the lower pressure, the air expands, but energy is needed to expand a volume of gas against the surrounding pressure; this energy comes from the motion of individual molecules of gas. In consequence, the molecular motion decreases, and with it the temperature, which is just a measure of the average speed of molecules. This decrease of temperature with pressure is quantified by the first law of thermodynamics. In our atmosphere, rising air cools about 1°C (1.8°F) for each 100 m (330 ft) it ascends.

The cooling of ascending air eventually causes water vapor within it to condense. Normally, individual molecules of water rebound off each other when they collide. But if their concentration is great enough, or if it is cold enough, they succeed in clustering into drops of liquid water (or ice, if it is very cold). This process is

called condensation. In reality, the water molecules collect on small particles, such as dust or sea salt, and form very small droplets that we see as cloud. These droplets are so small that they are effectively carried along in the updraft.

Thus when a warm, humid current of air has risen high enough, it becomes cold enough for the water vapor to condense into very small droplets, which we see as the base of the cloud. This condensation heats the air, so that as it continues to rise, it does not cool as fast is it did below cloud base. Just how fast it cools depends on how much water vapor is in the air, but initially, the rate of cooling can be as little as one-third of the rate below cloud base. This helps keep the air in the cloud warmer than its environment.

After twenty minutes or so, the cloud has grown so high that water vapor begins to condense into ice crystals. The updraft finally encounters the tropopause, above which the temperature of the environment no longer decreases rapidly with altitude, and the air inside the cloud can no longer remain warmer than its environment. The updraft stops abruptly, and the air and ice crystals within it spread out at or near the tropopause, forming an anvil cloud, an example of which is shown in Figure 8.3. At the same time, the myriad tiny water droplets and ice crystals that formed in the updraft begin to collide with one another, eventually forming drops, snowflakes, or hail big enough to fall at appreciable speeds. Some of this precipitation falls outside the cloud itself and partially re-evaporates in the dry air, cooling it and forming downdrafts. The storm has entered its mature stage (middle panel of Figure 6.1).

At about this time, the first rain reaches the surface. (Some of this rain forms in the lower part of the cloud, where the temperature is above freezing, and some of it comes from melting snow and hail that formed high up in the cloud.) The cold, heavy air, chilled by the evaporation of falling rain, forms a dome at the earth's surface that then spreads outward, like a bowl of soup spilled on a kitchen floor. The leading edge of this spreading cold air is often very sharp and is referred to by meteorologists as a "gust front." As it passes, the wind picks up and switches abruptly, usually flowing away from the darkest part of the sky where the heaviest rain is falling, and the temperature drops, often quickly, providing temporary relief from the heat and humidity of a normal day in the Tropics.

Yet this bulldozer of spreading cold air sows the seeds of destruction of the storm that created it. As it spreads further from the cloud, it begins to cut off the supply of warm, moist air to the updraft. The storm enters its dissipation stage (right panel of Figure 6.1). The updraft weakens, first near its base and later higher up in the cloud. Some precipitation continues to form in the spreading anvil cloud and to fall to the earth as light rain; it may take several hours for the anvil finally to dissipate. The downdraft ceases, and the cold pool of air near the surface gradually warms up, owing to its contact with the surface and the fact that it is sinking and compressing. After several hours, conditions return to normal.

This type of convective shower or thunderstorm is an everyday feature of those parts of the Tropics that breed hurricanes. It is a vital part of the heat balance of the tropical atmosphere (see Chapter 4), carrying heat away from the ocean, which has absorbed it from the sun. They are as important to the functioning of the climate system as the cells in our bodies are to our lives. And just as living cells occasionally go out of control, causing cancer, tropical showers sometimes mysteriously gang up into the raging meteorological monsters we know as hurricanes. But why?

Ships in Distress off a Rocky Coast. Oil painting by Dutch artist Ludolf Backhuysen, 1667.

All round the Horizon black Clouds appear;
   A Storm is near:
Darkness Eclipseth the Sereener Sky,
   The Winds are high,
Making the Surface of the Ocean Show
Like mountains Lofty, and like Vallies Low.

The weighty Seas are rowled from the Deeps
   In mighty heaps,
And from the Rocks Foundations do arise
   To Kiss the Skies
Wave after Wave in Hills each other Crowds,
As if the Deeps resolv'd to Storm the Clouds.

How did the Surging Billows Fome and Rore
   Against the Shore

Threatning to bring the Land under their power
   And it Devour:
Those Liquid Mountains on the Clifts were hurld
As to a Chaos they would shake the World.

The Earth did Interpose the Prince of Light,
   'Twas Sable night:
All Darkness was but when the Lightnings fly
   And Light the Sky,
Night, Thunder, Lightning, Rain, and raging Wind,
To make a Storm had all their forces joyn'd.

   —Richard Steere (1643–1721),
     "On a Sea-Storm nigh the Coast"

7 France Gives Up La Floride, 1565

I hold it that a little rebellion, now and then,
is a good thing, and as necessary in the
political world as storms in the physical.

—Thomas Jefferson

Were it not for a hurricane, France, rather than Spain, might have successfully claimed Florida in the middle of the sixteenth century.

The 1500s were a time of great religious turmoil in France, with increasing persecution of French Protestants, known as Huguenots. One of the leaders of the Huguenots was Gaspard de Coligny, a French admiral. At the height of the turmoil, Coligny appointed the mariner Jean Ribaut to establish a New World asylum for Huguenots. Ribaut set sail in 1562, arriving on May 1 at the mouth of what is today known as the St. Johns River. Forty-nine years after Ponce de Leon had claimed "Pascua Florida" for Spain, Ribaut erected a stone monument emblazoned with the French coat of arms, declaring the territory the possession of the king of France. He then sailed north, establishing a colony he called Charlesfort (Parris Island), South Carolina, returning to Dieppe in July with the intention of sending settlers and aid back to Charlesfort. Finding the Huguenots and Catholics at war, Ribaut fled to England to seek ships and supplies for his New World colony, only to be imprisoned in the Tower of London, suspected of planning to steal English ships. When aid was not forthcoming, the 28 Charlesfort colonists mutinied against and killed their commander and, abandoning the settlement, returned to France in a ship of their own construction.

Meanwhile, Admiral Coligny sent Ribaut's lieutenant, René Goulaine de Laudonnière, to Florida with a group of colonists. Arriving at the mouth of the St. Johns River, and helped by the Timuqua chief Athore, they built a fortified settlement named Fort Caroline, in honor of the French king Charles IX. Things did not go well at Fort Caroline. The French had great difficulty adapting to the local climate and swampy terrain, and after less than a year, they abandoned the settlement and returned to France. By then, the English had released Ribaut, who was then sent by Coligny to Fort Caroline with seven ships and about six hundred new colonists. He met Laudonnière's party just as the latter was setting off. At the same time, a fleet under the command of Admiral Pedro Menéndez de Avilés was on its way with orders to eliminate the French at Fort Caroline and reclaim the territory for Spain. Menéndez had set sail from Spain with 11 ships, but his fleet was decimated by a hurricane. Only five of the ships survived. After briefly confronting the French fleet at Fort Caroline, Menéndez regrouped at St. Augustine.

Ribaut decided to attack St. Augustine. On September 19, 1565, he set sail with 400 men for the attack, leaving about 250 soldiers at Fort Caroline. The Spanish were badly outnumbered and surely would have succumbed to the French invasion were it not for a fateful intervention.

While on route, the French fleet was overtaken by a strong hurricane and wrecked some distance south of St. Augustine. Remarkably, there were few casualties. Suspecting (correctly) that Menéndez would try to attack the thinly defended Ft. Caroline, Ribaut and his battered men tried to make their way there. Outside St. Augustine, they encountered a detachment from Menéndez's force. Menéndez tricked the French into surrendering, promising to spare their lives, then proceeded to butcher all of them, including Ribaut. None of Ribaut's men knew that Menéndez had already attacked and killed most of the inhabitants of Fort Caroline. The Spanish later built Fort Matanzas (Spanish for "massacre") at the site of the murder of Ribaut and his men.

It is an exercise in "virtual history" to imagine what might have transpired if Ribaut's invasion of St. Augustine had not been thwarted by a hurricane. Perhaps the Spanish would eventually have prevailed in Florida anyway. On the other hand, it is just possible that millions of tourists would now be flocking to La Floride to visit Le Monde Disney.

▲

The Hurricane. Oil painting by American artist Paul A. L. Hall, ca. 1966.

Sleep at noon. Window blind
rattle and bang. Pay no mind.
Door go jump like somebody coming:
let him come. Tin roof drumming:
drum away—she's drummed before.
Blinds blow loose: unlatch the door.
Look up sky through the machineel:
black show through like a hole in your heel.
Look down shore at the old canoe:
rag-a-tag sea turn white, turn blue,
kick up dust in the lee of the reef,
wallop around like a loblolly leaf.
Let her wallop—who's afraid?
Gale from the north-east: just the Trade…

And that's when you hear it: far and high—
sea-birds screaming down the sky
high and far like screaming leaves;
tree-branch slams across the eaves;
rain like pebbles on the ground…

and the sea turns white and the wind goes round.

　　　—Archibald MacLeish (1892–1982),
　　　　"Hurricane"

8   The Trade Winds

For they have sown the wind, and they shall reap the whirlwind.

—Hosea 8:7

During his voyages to the New World, Columbus found that as soon as he was south of about 30° north latitude, the variable and stormy winds of Europe gave way to steady easterlies and northeasterlies. These made his westbound journeys fast and comfortable, but greatly lengthened his return trips.

By the seventeenth century, it had become well established that easterly winds prevail in nearly all of the Tropics (which is why global circumnavigations were almost always sailed from east to west). In 1686, the great British scientist Sir Edmund Halley (after whom the comet was named) collected wind data from ships' logs and used them to construct a map of tropical surface winds (Figure 8.1). This is thought to be the first meteorological map ever published.

Halley wondered what made the winds

Figure 8.1: Map of tropical airflow constructed by Edmund Halley in 1686. This is thought to be the first weather map ever published.

▼

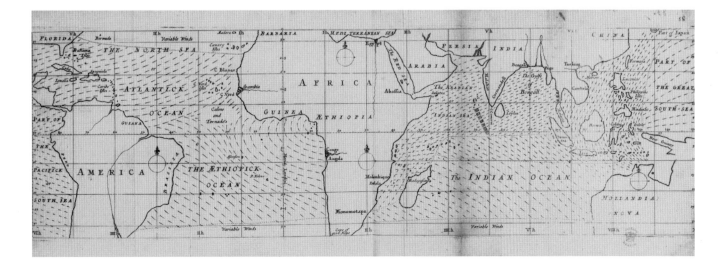

blow systematically from east to west. In a 1685 paper entitled "An Historical Account of the Trade Winds, and Monsoons, observable in the Seas between and near the Tropicks, with an attempt to assign the Physical Cause of the Said Winds," he conjectured that air tends to "follow the sun" as it moves from east to west across the sky. This effect is not complete nonsense: laboratory experiments confirm that moving a heat source around the rim of a pan of water does give rise to a motion that follows the heat source, but the effect is far too weak to explain the trade winds.

A nearly correct explanation was supplied some 50 years later by another Englishman and member of the Royal Society, George Hadley. In his 1735 masterpiece *Concerning the Cause of the General Trade-Winds*, Hadley argued that

equatorial regions are necessarily hotter than higher latitudes, because sunlight strikes the earth more nearly at right angles near the equator, so that each unit area of the surface intercepts more sunlight. Hadley recognized that warm air is less dense than cold air at the same pressure, and he reasoned that the warmer air near the equator would rise and be replaced by cooler air flowing equatorward from higher latitudes. This would imply northerly winds north of the equator and southerlies south of the equator. To explain why the winds in fact blow more nearly from the east, Hadley attempted to apply Isaac Newton's first law of motion. Hadley's reasoning is illustrated in Figure 8.2.

Hadley had calculated that to an observer stationary in space, an object fixed to the equator would be moving from west to east at over

Figure 8.3: A tropical thundershower, with an anvil cloud at its top. The anvil is composed of tiny ice crystals.

▲

450 m/s (1,000 mph), owing to the earth's rotation. But an object fixed at 30° north latitude would be moving at only 400 m/s (900 mph). Hadley thought that if this object started to move toward the equator, it would preserve its eastward velocity of 400 m/s, so that by the time it got to the equator, it would still be moving eastward at 400 m/s, or more than 45 m/s (100 mph) *slower* than the earth underneath it. Thus, to an observer at the equator, the object would be moving from east to west at more than 45 m/s. Hadley reasoned that the same thing would happen to air moving from higher toward lower latitudes.

Although he had the right idea, Hadley had misapplied Newton's first law. In a rotating system, it is not linear momentum that is conserved, as Hadley supposed, but *angular momentum*, which is the product of the west-to-east

velocity and the distance from the axis of rotation. When an object moves from 30° north latitude toward the equator, it moves further away from the earth's axis of rotation. To preserve its angular momentum, the object actually has to slow down. This means that when it reaches the equator, it is traveling somewhat more slowly than 400 m/s (900 mph)—actually about 350 m/s (780 mph). Thus, relative to an observer on the equator, the object is actually moving from east to west at more than 110 m/s (250 mph). Fortunately, friction between the moving air and the earth's surface slows the winds way down, to the more modest 5–10 m/s (10–25 mph) of the trades.

Although he missed some of the details, Hadley was remarkably prescient in his view of the physics of the trade winds. Day after day, air near the surface does indeed flow toward the

Figure 8.4: Trade cumulus clouds are a trademark of the tropical skyscape.

▲

equator in both hemispheres, curving toward the west as it does so, forming the northeast trades in the Northern Hemisphere and the southeast trades south of the equator. These two streams of air converge into a surprisingly narrow zone called the Inter-Tropical Convergence Zone, or ITCZ. Here the converging air ascends in towering cumulonimbus clouds, producing very heavy tropical downpours as the abundant water vapor of the tropical air condenses in the rising, cooling air stream (see Chapter 6). By the time air flows out of the tops of these clouds, the tallest thunderstorms on earth, it has reached a height of more than 16 km (10 mi; well above the maximum cruising altitude of commercial aircraft) and cooled to temperatures below -70°C (-100°F). What little water is left is mostly in the form of tiny ice crystals that make up the anvil clouds often seen at the tops of thunderstorms (Figure 8.3).

Air flowing out of the tops of the cumulonimbi moves poleward, conserving its angular momentum. By the time it reaches 20° latitude in either hemisphere, it is moving eastward at over 45 m/s (100 mph). Because it is far from the surface, there is little friction to slow it down. Somewhere in the 20° to 30° latitude belt, the air begins to sink back toward the surface, gradually losing its eastward momentum by exchanging it with air at higher latitudes. The air then repeats the cycle, flowing back toward the ITCZ near the surface.

This circulation of air, flowing equatorward at the surface, rising in the ITCZ, flowing back to the subtropics at high altitude, and sinking back toward the surface, is called a Hadley Cell, after its discoverer.

The air that sinks in the subtropics is bone dry, having lost almost all of its water in downpours as it ascended in the deep cumulonimbi of the ITCZ. But air in contact with the relatively cool sea surface is moistened by evaporation of seawater. The warm, dry descending air meets this boundary layer of cool, moist air at a sharp temperature inversion known as the trade inversion, usually about 1.5 km (6,000 ft) above the sea surface. (A temperature inversion is a region of the atmosphere where temperature increases with height; usually temperature falls with elevation.) Above the trade inversion, there are no clouds save for the occasional wisp of high cirrus, ghostly remains of the tops of distant cumulonimbi. Below the trade inversion, the sky is filled with small, pretty, ever-changing cumulus clouds known as trade cumuli. Trade cumuli are nearly ubiquitous over the tropical seas, but are usually too small to produce rain. No idyllic travel agency poster of tropical resorts would be complete without them (Figure 8.4). They serve to remoisten the very dry air descending through the trade inversion.

The actual distribution of winds is not quite as simple as I have suggested. First, the ITCZ moves back and forth across the equator at many longitudes, favoring the summer hemisphere. The winter-hemisphere Hadley cell is usually much stronger than the summer-hemisphere Hadley cell.

In addition, the winds vary with longitude, owing to the presence of continents and ocean currents. Even Halley's map (Figure 8.1) hints at this, but it can be clearly seen in maps of surface winds, such as that shown in Figure 8.5. This map was made using a satellite-borne instrument called a scatterometer. As the satellite passes over the ocean, it emits a series of pulses of radiation, directed at an angle down toward the ocean surface. These radiation pulses scatter off the tiny capillary waves that we see as ripples whenever the wind blows over water. By looking at these waves from several different angles as the satellite passes by, and by measuring how much radiation returns to the satellite, one can estimate their height and orientation, and determine the direction and speed of the wind. This technological miracle allows us to estimate the wind speed and direction over almost all of the world's oceans many times each day.

The map of surface winds over the Pacific in Figure 8.5 clearly shows the northeast and southeast trades converging at a wavy line (the ITCZ) centered around 10° north latitude. (The winds flow along the white lines on the map, and their speed is given by the color scale at right.) But in the far eastern Pacific, the winds are more northerly north of the ITCZ and more southerly south of it, with little westward component. There are strong *westerly* winds to the east of the Philippines and in the South China Sea, and the airflow in the western North Pacific is dominated by two cyclonic (counterclockwise) vortices: a typhoon east of Taiwan and a lesser tropical disturbance well east of the Philippines.

The tropical climate may be thought of as a radiative-convective equilibrium state with a Hadley Circulation superimposed. Sunlight is absorbed by the ocean, which loses heat mostly

Figure 8.5: Map showing winds over the Pacific Ocean, as measured by a scattero-meter aboard NASA's QuickSCAT satellite. The direction of the wind is shown by the white curves with arrows, and the colors show the wind speed as given by the bar scale at right. The scatterometer transmits pulses of microwave radiation that are backscattered from capillary waves (ripples) on the sea surface. By measuring how much radiation is returned, one can estimate the height of the capillary waves, and this is proportional to the wind speed. As the satellite passes overhead, it transmits pulses at various angles to the sea surface; by measuring how much radiation is returned as a function of the angle, one can estimate the orientation of the capillary waves, and this gives the wind direction.

▶

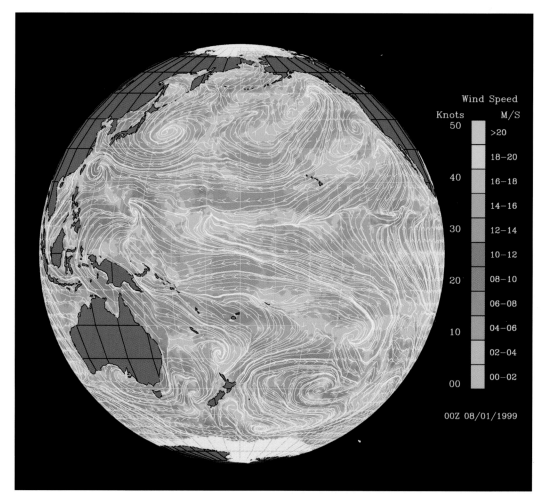

by evaporation of water. This water is carried upward in deep cumulonimbus clouds, where it precipitates out as rain. As altitude increases in the tropical atmosphere, more and more heat is carried upward by infrared radiation, and less and less by convection, until at the tropical tropopause, 16 km (10 mi) above the surface, all the heat is carried by radiation.

Thanks to the Hadley Circulation, the deep convection is not scattered uniformly through the Tropics, but instead it is concentrated in a few regions, such as the ITCZ and the "monsoon trough," an area of low pressure often found over the warmest ocean water of the tropical western Pacific. Elsewhere, air is slowly descending, and the radiative cooling of the atmosphere is balanced not so much by heating in tall convective clouds, but by the compression of the air as it sinks toward the surface. In such places, the sky is often filled with "trade cumuli," which are too small to rain. But it is the rising branch of the Hadley Circulation, haunted by intense rain showers and thunderstorms, that serves as the breeding ground of hurricanes.

▲

Miranda, The Tempest.
Oil painting by British
artist John W. Waterhouse,
1916.

I f by your art, my dearest father, you have

Put the wild waters in this roar, allay them.

The sky, it seems, would pour down stinking pitch

But that the sea, mounting to th' welkin's cheek,

Dashes the fire out. O, I have suffered

With those that I saw suffer! a brave vessel

(Who had no doubt some noble creature in her)

Dashed all to pieces! O, the cry did knock

Against my very heart! Poor souls, they perished!

Had I been any god of power, I would

Have sunk the sea within the earth or ere

It should the good ship so have swallowed and

The fraughting souls within her.

—Miranda, to her father Prospero, in Shakespeare (1564–1616),
The Tempest, Act I, Scene ii

9 The Tempest

I had as little hope as desire of life in the storm,
and in this, it went beyond my will.

— William Strachey, onboard the *Sea Venture*

Hurricanes often entrain seeds, insects, and small birds into their immense circulations, carrying them far from their native habitats to distant lands, where they may take root and colonize. Here is the story of how one hurricane blew a ship off course, depositing its human cargo on an isolated tropical island where they founded a colony that persists to this day. This story almost certainly served as the inspiration for William Shakespeare's last play.

On June 2, 1609, a fleet of seven tall ships, two with pinnaces in tow, sailed from the English port of Plymouth, bound for America. With their cargo of nearly six hundred passengers, they had been sent by the Virginia Company of London to fortify the Jamestown settlement. The lead ship, the three-hundred-ton *Sea Venture*, was the largest in the fleet and carried Sir Thomas Gates, the newly appointed governor of the colony, and Sir George Somers, admiral of the Virginia Company.

The first few days of the voyage were uneventful, but as the fleet drew near the Azores, a hurricane scattered the ships. They continued on toward the west, lost to the sight of each other. All but one of the original ships made it to Jamestown; the *Sea Venture* never arrived and was presumed lost. What actually happened aboard the *Sea Venture* is described in vivid detail in a letter sent back to England by William Strachey, who had been appointed secretary to the deputy governor of Virginia. Here are excerpts from his letter:

> S. James his day, July 24, being Monday (preparing for no less all the black night before), the clouds gathering thick upon us, and the winds singing, and whistling most unusually, which made us to cast off our Pinnace…a dreadful storm and hideous began to blow from out the North-east, which swelling, and roaring as it were by fits, some hours with more violence then others, at length did beat all light from heaven; which like a hell of darkness turned black upon us,

so much the more fuller or horror, as in such cases horror and fear use to overrun the troubled, and overmastered senses of all, which (taken up with amazement), the ears lay so sensible to the terrible cries, and murmurs of the winds, and distraction of our company, as who was most armed, and best prepared, was not a little shaken....

For four and twenty hours the storm in a restless tumult, had blown so exceedingly, as we could not apprehend in our imaginations any possibility of greater violence, yet did we still find it, not only more terrible, but more constant, fury added to fury, and one storm urging a second more outrageous than the former; whether it so wrought upon our fears, or indeed met with new forces: Sometimes strikes in our ship amongst women, and passengers, not used to such hurly and discomforts, made us look one upon the other with troubled hearts, and panting bosoms: our clamors drowned in the winds, and the winds in thunder. Prayers might well be in the heart and lips, but drowned in the outcries of the officers: nothing heard that could give comfort, nothing seen that might encourage hope. The Sea swelled above the clouds, and gave battle unto heaven. It could not be said to rain; the waters like whole rivers did flood in the air. And this I did still observe, that whereas upon the land when a storm hath poured itself forth once in drifts of rain, the wind as beaten down, and vanquished therewith, not long after endureth: here the glut of water (as if throttling the wind ere while) was no sooner a little emptied and qualified, but instantly the winds (as having gotten their mouths now free, and at liberty) spake more loud, and grew more tumultuous, and malignant.

Howbeit this was not all; it pleased God to bring a greater affliction yet upon us; for in the beginning of the storm we had received likewise a mighty leak. And the ship in every joint almost, having spewed out her oakum, before we were aware was grown five foot suddenly deep with water above her ballast, and we almost drowned within, whiles we sat looking when to perish from above. This imparting no less terror than danger, ran through the whole ship with much fright and amazement, startled and turned the blood, and took down the braves of the most hardy Mariner of them all, insomuch as he that before happily felt not the sorrow of others, now began to sorrow for himself, when he saw such a pond of water so suddenly broken in and which he knew could not (without present avoiding) but instantly sink him. The Lord knoweth, I had as little hope as desire of life in the storm, and in this, it went beyond my will; yet we did, either because so dear are a few lingering hours of life in all mankind, or that our Christian knowledges taught us, how much we owed to the rites of Nature, as bound, not to be false to our selves, or to neglect the means of our own preservation; the most despaireful things amongst men, being matters of no wonder no moment with Him, who is the rich fountain and admirable essence of all mercy.

On the Thursday night Sir George Summers being upon the watch, had an apparition of a little round light, like a faint star, trembling, and streaming along with a sparkling blaze, half the height upon the mainmast, and shooting sometimes from shroud to shroud, tempting to settle as it were upon any of the four shrouds.... The superstitious seamen make many constructions of this Sea-fire, which nevertheless is usual in storms: the same which the Grecians were wont in the Mediterranean to call Castor and Pollux, of which, if one only appeared without the other, they took it for an evil sign of great tempest.

But see the goodness and sweet introduction of better hope, by our merciful God given unto us. Sir George Summers, when no man dreamed of such happiness, had discovered, and cried Land. We found it to be the dangerous and dreaded island, or rather islands, of the

Bermuda. And that the rather, because they be so terrible to all that ever touched on them, and such tempests, thunders and other fearful objects are seen and heard about them that they be called commonly, the devils islands, and are feared and voided of all sea travelers alive, above any other place in this world. Yet it had pleased our merciful God, to make even this hideous and hated place both the place of our safety, and means of our deliverance.

And hereby I hope to deliver the world from a foul and general error: it being counted of most, that they can be no habitation for men, but rather given over to devils and spirits; whereas indeed we find them now by experience, as habitable and commodious as most countries of the same climate and situations: insomuch as if the entrance into them were as easy as the place itself is contenting, it had long ere this been inhabited, as well as other islands. Thus shall we make it appear that truth is the daughter of time, and that men ought to deny every thing which is not subject to their own sense.[5]

While the fleet sailed west from the Azores, a hurricane made its way north from the West Indies. The storm struck the *Sea Venture* on Monday, July 24, throwing her into a state of utter chaos. It took eight men to hold the helm, even under bare poles, and a great deal of cargo…luggage, food and drink, goods of all kinds…was heaved overboard to lighten ship. But she soon sprang a leak, and seawater flowed inexorably into her hold, filling it to a depth of ten feet. Everyone on board worked to save his life, manning the pumps and attempting to stem the flow of water by stuffing salt beef and anything else they could find into the leaks. Sir George did everything in his power to save the ship, remaining on deck for three days and nights while coordinating the desperate efforts to keep the *Sea Venture* afloat. But by Friday morning, the twenty-eighth, Somers and his charges faced the awful truth that they were sinking in mid ocean. Most of the crew gave up hope, falling asleep were they could, exhausted from their relentless but futile efforts.

Just then, as if by miracle, a rocky coast loomed ahead. It was the *Ya de Demonios*…"Islands of Devils"…first charted in 1511 by the Spanish explorer Juan Bermudez. While the crew pumped with renewed vigor, Somers skillfully navigated the foundering *Sea Venture* onto a reef about a half mile to the lee of the main island, and used the ship's long boat to ferry the crew and passengers ashore. Remarkably, everyone was saved and, as Strachey notes in the last paragraph quoted here, they found the island considerably less inhospitable than its reputation had led them to expect. Although the ship was rebuilt and sailed on to Jamestown, some of its crew and passengers elected to stay on the island, and Somers himself eventually returned and remained there until his death. In 1615, the Bermuda Company was formed and granted a charter by King James. Each year, July 28 is celebrated on the Island as Somers Day.

In 1610, Strachey's dramatic account of the storm and the heroic efforts to save the *Sea Venture* reached England, where it was quickly circulated among Strachey's friends and many of those interested in the American colonies. Among Strachey's acquaintances was William Shakespeare, who almost certainly read the account and used it as the basis of his last play, *The Tempest*, thought to have been penned from late 1610 to 1611. Although not set in Bermuda, the play recounts events of

---

[5] From William Strachey's "True repertory of the wreck and redemption of Sir Thomas Gates, July 15, 1610" (pub. 1625, in *Purchas His Pilgrimes*, Part 4, Book 9, Chapter 6, but almost certainly circulating in manuscript before then).

uncanny similarity to those described by Strachey, using some phrases identical to those in Strachey's letter. The magical figure Ariel may very well have been inspired by Strachey's account of the electrical phenomenon known as St. Elmo's Fire, described in the fourth paragraph quoted earlier. Here is how Shakespeare has Ariel describe her luminous doings:

> *I boarded the king's ship; now on the beak,*
> *Now in the waist, the deck, in every cabin,*
> *I flamed amazement: sometime I'd divide,*
> *And burn in many places; on the topmast,*
> *The yards and bowsprit, would I flame distinctly,*
> *Then meet and join. Jove's lightnings, the precursors*
> *O' the dreadful thunder-claps, more momentary*
> *And sight-outrunning were not; the fire and cracks*
> *Of sulphurous roaring the most mighty Neptune*
> *Seem to besiege and make his bold waves tremble,*
> *Yea, his dread trident shake.*

Many Shakespeare scholars hold that the similarity between *The Tempest* and Strachey's account of the fate of the *Sea Venture* is the final nail in the coffin of the theory that Shakespeare was actually Edward de Vere, the seventeenth earl of Oxford, who died in 1604 and thus could not have heard Strachey's remarkable story.

▲

Inundación en Guatemala
[Hurricane Mitch].
Oil painting by Diego
Isaias Hernandez Mendez,
Tz'utujil Maya artist from
Guatemala, 2001.

O God! when thou
Dost scare the world with tempests, set on fire
The heavens with falling thunderbolts, or fill,
With all the waters of the firmament,
The swift dark whirlwind that uproots the woods
And drowns the villages; when, at thy call,
Uprises the great deep and throws himself
Upon the continent, and overwhelms
Its cities—who forgets not, at the sight
Of these tremendous tokens of thy power,
His pride, and lays his strifes and follies by?

   —William Cullen Bryant (1794–1878), from "A Forest Hymn"

10 Nature's Steam Engine

*C'est à la chaleur que doivent être attribués les grands mouvements qui
frappent nos regards sur la terre; c'est à elle que sont dues les agitations
de l'atmosphère, l'ascension des nuages, la chute des pluies et des autres
météores, les courants d'eau qui sillonnent la surface du globe et dont
l'homme est parvenu à employer pour son usage une faible partie; enfin les
tremblements de terre, les éruptions volcanique reconnaissent aussi pour
cause le chaleur.*

—Nicolas Léonard Sadi Carnot[6]

Thus the great nineteenth-century French scientist Carnot attributed to the flow of heat the grand motions of the atmosphere, the flow of rivers, and the forces inside the earth that drive earthquakes and volcanoes. Physicists such as Carnot were motivated, in part, by a desire to understand the theory underlying the operation of steam engines, and they were particularly concerned to determine how efficient such engines can be made. Carnot showed that for any engine to convert heat energy into mechanical energy, heat has to flow from a high-temperature reservoir to a low-temperature reservoir, and he suggested that the fraction of heat energy that can be converted is proportional to the difference between the temperatures of the two reservoirs.

Carnot's other great contribution to thermodynamics was his demonstration that a particular kind of heat engine maximizes the conversion of heat energy to mechanical energy. This ideal heat engine works on a particular thermodynamic cycle in which a working substance is transformed in four steps, ending up in the same state that it began. He illustrated this cycle by considering a gas in a cylinder contained by a piston (Figure 10.1):

Step 1: The gas is heated and at the same time, the piston rises, reducing the pressure on the gas. When pressure falls, the gas cools, but in this case, just enough heat is added to keep the temperature of the gas constant. This is called *isothermal expansion.*

Step 2: The heating is turned off, but the piston continues to rise, further reducing the pressure on the gas. As the pressure is reduced, the gas cools. This is called *adiabatic expansion.*

---

[6] "It is to heat we should attribute the great movements that appear to us on earth; to it are due the agitations of the atmosphere, the rising of clouds, the fall of rain and other meteors, the water currents that furrow the surface of the globe and of which man has managed to turn a small part to his use; and finally, earthquakes and volcanic eruptions are also caused by heat." *Reflexions sur la Puissance Motrice du Feu,* Paris, 1824.

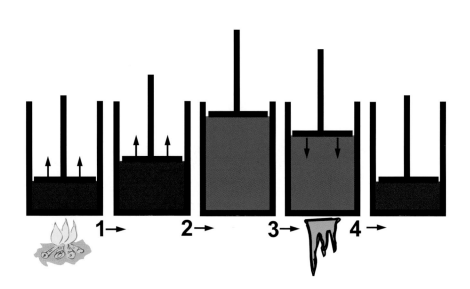

("Adiabatic" means "without addition of heat.")

Step 3: In the reverse of step 1, the gas is cooled and at the same time, the piston falls, increasing the pressure on the gas at just such a rate as to keep its temperature constant. This is called *isothermal compression*, and the gas loses heat during this step.

Step 4: In this last step, the cold source is removed, and the gas is compressed until it warms to the temperature it had at the start of the cycle. This is called *adiabatic compression*.

Carnot showed that this cycle, which now bears his name, does work on whatever is attached to the piston. So, for example, if the piston is connected to a crankshaft, some of the heat energy (supplied by the fire in this example) is used to accelerate the shaft or, if there is enough friction somewhere in the system, to keep the shaft turning against friction. As noted earlier, he also showed that the fraction of energy supplied by the heat source that is ultimately used to do this work is proportional to the difference between the temperatures of the substance (gas, in this case) in steps 1 and 3. The greater this temperature difference, the more work can be done for the same input of heat energy. This relation can be expressed mathematically. Let W be the mechanical work that can be done by the engine, and Q the rate of heat input. Also let T_{hot} be the absolute temperature[7] of the substance at the beginning of the cycle, and let T_{cold} be the temperature at the end of step 2. Then

$$W = Q\left(\frac{T_{hot} - T_{cold}}{T_{hot}}\right). \qquad (10.1)$$

An automobile engine is an example of a heat engine. Heat is put into the engine when a spark plug ignites the mixture of gasoline and air inside each cylinder. This happens when the piston is nearly at the end of the cylinder, and the mixture is at maximum compression and maximum temperature. (In a diesel engine, the gases get so hot from compression that they ignite

---

[7]The absolute temperature is measured in degrees Kelvin, where 0 Kelvin is "absolute zero." This is the coldest temperature any substance can attain, at which all molecular motions cease. 0 Kelvin is equal to -273 C or -459 F.

Figure 10.2: The energy cycle of a mature hurricane. Air spirals inward close to the sea surface, between points A and B, acquiring heat from the ocean by evaporation of seawater. Air then ascends in the eyewall, from B to C, without acquiring or losing heat other than that produced when water vapor condenses. Between C and D, the air loses the heat it originally acquired from the ocean. Finally, between D and A, the air returns to its starting point. In a real hurricane, the energy cycle is open because hurricanes continuously exchange air with their environment (see text). The colors show a measure of the air's heat content, with warm colors corresponding to high heat content.

▶

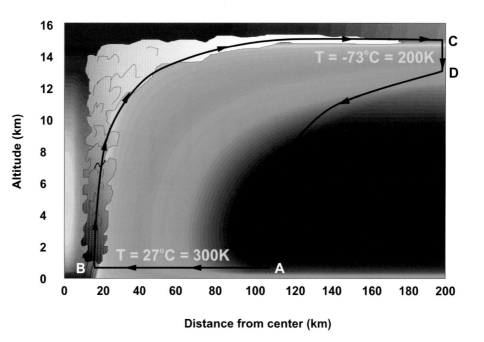

without a spark plug.) After the miniature explosion takes place, the gases in the cylinder expand and cool, and the piston is forced rapidly downward. Heat is then taken away from the cylinder by the circulation of cool water (or air, if it's an old VW Beetle) through the engine and ultimately exchanged to the atmosphere through the car's radiator.

The auto engine departs in several respects from the ideal heat engine first described by Carnot. But it turns out that a hurricane is an almost perfect example of a Carnot heat engine.

In a hurricane, the working substance is not just air, but a mixture of moist air, water droplets, and ice crystals. (We could choose dry air as the working substance, but then we would have to take into account the enormous conversion of energy that takes place when water vapor condenses into liquid water drops or ice crystals. By choosing moist air and condensed water as the working fluid, we can mostly avoid dealing with phase changes of water.) The heat cycle of the hurricane is shown in Figure 10.2.

This diagram shows a cross section through an idealized hurricane. The central axis of the storm is on the left side of the diagram. The wind speeds of the storm are assumed not to be changing in time, and we neglect any variations in the properties of the air as one moves around the axis of the storm. Let's follow a sample of air that begins at point A near the sea surface, a hundred kilometers or so from the storm center. The air begins to spiral in toward the eyewall at point B. The storm center is an area of relatively low pressure, so the pressure on the air decreases as it travels from A to B. But it is always in contact with the sea surface, which acts as an almost unlimited heat reservoir, so its temperature is approximately constant. It is during this leg that enormous quantities of heat are added to the air. Leg A–B is the firebox of the hurricane.

At first it does not seem obvious that air is heated as it flows from A to B. But remember that air flowing from high to low pressure would cool were no heat added to it. Here, heat flowing from the sea into the air keeps it at nearly constant temperature. But there is a much more important source of heat: the enormous flow of energy that occurs when seawater evaporates into the inflowing air. Just as in a steam engine, the heat from the ocean is used mostly to evaporate water. Evaporation of water is a very efficient way to transfer heat from one body to another. That is why you feel cold when you are wet, especially if it is windy and/or dry: evaporation is taking heat from your body.

Thus the addition of heat in leg A–B shows up mostly as an increase in the humidity of the air. This form of heat is called *latent heat. Evaporation of seawater into the inflowing air is by far the most important source of heat driving the hurricane.* And while evaporation occurs over most of the area covered by the storm, the part of the evaporation that is actually effective in driving the storm occurs very near the hurricane's eyewall, where the winds are strongest. When hurricanes make landfall, they quickly die because they are cut off from their oceanic energy source.

There is one further effect on the energy of hurricanes. The terrific winds blowing across the surface near the eyewall are constantly being dissipated by friction. Just as you produce heat when you rub your hands together, this frictional dissipation heats the air near the surface. We shall return to this point in a moment.

The colors in Figure 10.2 show the entropy of the moist air; very loosely, this measures the total heat content of the air (including the latent component). Note that as air approaches the eyewall, its entropy increases rapidly, reflecting the large input of heat from the ocean.

Now at point B, the air turns and flows upward through the towering cumulonimbus clouds that make up the hurricane's eyewall. It is here that the latent heat is converted into sensible heat as water vapor condenses. But it is important to note that although there is enormous conversion between these two forms of heat, the *total* heat content (entropy, actually) remains approximately constant along this leg. (Note that the colors do not change along leg B–C.) As the air flows upward, the pressure on it decreases very rapidly. What we have here is an example of adiabatic expansion.

In the real world, the heat absorbed from the ocean and shot upwards through the hurricane's eyewall is expelled into the distant environment. But in computer models, we can place walls around the outside of the storm, forcing the air to return to the surface and thereby closing the loop. This makes it a little easier to describe the thermodynamic cycle of the storm.

By the time the air reaches point C in the high troposphere or lower stratosphere, some 12 to 18 km above the sea surface, the adiabatic expansion has lowered its temperature to a value close to that of the undisturbed upper tropical atmosphere, around -70°C or about 200 Kelvin. The air remains at approximately this temperature as it sinks down toward the tropopause (point D). Although the air is undergoing compression, it is losing heat by electromagnetic radiation to space (see Chapter 4). This leg of the cycle is very nearly one of isothermal compression.

Finally, the air sinks from point D back to the starting point A. In reality, the air is losing heat by radiation to space, just as in leg C–D, but it turns out that the amount of heat lost is almost equivalent to the amount of heat that would have been lost if the rainwater, instead of falling out of the storm, had remained in the air, evaporating as the air descends. This latter is, once again, just a conversion between two different forms of heat that preserves the total heat content. Thus leg D–A is very nearly one of adiabatic compression.

The thermodynamic cycle of a mature hurricane is almost exactly like the idealized cycle envisioned by Carnot. The hurricane would be a perfect steam engine but for one interesting feature.

In Carnot's cycle, as in real steam engines, the mechanical energy produced by the engine is used to do something outside the engine itself, like power a locomotive. If the locomotive is moving at constant speed, then there is an equilibrium between the mechanical energy production by the engine and the frictional dissipation of energy of the wheels on the tracks, the air moving past the train, and all the moving parts of the engine and drivetrain. This frictional dissipation turns the kinetic energy into heat, which is lost to the environment.

The mechanical energy produced by the hurricane's heat engine shows up as the energy of the winds. But in a mature hurricane, almost all of the frictional dissipation occurs in the inflow layer. Thus the power of the winds is converted back into heat, which then flows back into the system at the high temperature reservoir where heat energy is being injected in the first place. Thus, unlike in the locomotive, some of what would have been wasted heat energy is recycled back into the front end of the heat engine. This recycling of waste heat makes hurricanes somewhat more powerful than they would be otherwise.

We can use our understanding of the hurricane's Carnot cycle to estimate how strong the winds can get, assuming that the sea surface temperature remains unaffected by the storm. First,

the rate at which kinetic energy is dissipated in the atmosphere near the surface is given by

$$D \approx C_D \rho V^3, \qquad (10.2)$$

where D is the rate of energy dissipation per unit area of the surface, ρ is the density of the air, V is the wind speed and C_D is a number called the *drag coefficient*, which is a property of the surface itself and increases with the roughness of the surface. Thus, the dissipation of kinetic energy increases rapidly with wind speed. Now the rate at which heat is added to the inflowing air depends on two processes: evaporation from the ocean and the conversion of kinetic energy back into heat. The total rate of heat input per unit area of the earth's surface is given by

$$Q \approx C_K \rho V E + C_D \rho V^3, \qquad (10.3)$$

where the new symbols are C_K, the *enthalpy exchange coefficient*, and E, the evaporative potential of the sea surface. The first is a number like the drag coefficient, but it measures how quickly heat flows across the air-sea interface. The second is a measure of the potential for transferring heat from the ocean to the atmosphere and is a function both of the air-sea temperature difference (which is usually a small effect in the Tropics) and the relative humidity of the air near the surface. The lower the relative humidity, the more water can evaporate into the air and the bigger the value of E. In fact, E is a measure of the air-sea thermodynamic disequilibrium and is a direct consequence of the greenhouse effect (see Chapter 4). The more greenhouse gases (and clouds) in the atmosphere, the less heat can escape from the ocean by means of radiation and the more heat has to get out by evaporation, thus the larger the value of E.

According to equation (10.1), the amount

of work produced by the hurricane's Carnot cycle is just the heat input, given by equation (10.3), multiplied by the thermodynamic efficiency, $\left(T_{hot} - T_{cold} \right) / T_{hot}$:

$$W \approx (C_K \rho VE + C_D \rho V^3) \text{ x } \left| \frac{T_{hot} - T_{cold}}{T_{hot}} \right| . \quad (10.4)$$

Equating the work given by equation (10.4) to the dissipation of kinetic energy given by equation (10.2) and doing a little algebra gives a formula for the maximum wind speed in a hurricane:

$$V_{max} \approx \sqrt{ \left| \frac{T_{hot} - T_{cold}}{T_{cold}} \right| E } . \quad (10.5)$$

Although we have derived this equation by approximate arguments, the same equation has been derived exactly using a different approach published in the professional literature. Note that now T_{cold} rather than T_{hot} appears in the bottom part of the fraction in equation (10.5). This is a consequence of allowing the feedback effect of dissipative heating, and it makes V_{max} larger than it would otherwise be. We shall refer to the maximum wind given by equation (10.5) as the *potential intensity* of the hurricane. To calculate it, we need to know the ratio of the surface exchange coefficients, the modified thermodynamic efficiency, and E. To get this last number, all we need to know is the sea surface temperature, the temperature and humidity of the air just above the surface, and the surface pressure at the location of maximum winds. Unfortunately, we do not know in advance what this surface pressure will be; it depends on the maximum wind speed itself. The solution to this problem is to make a guess at the surface pressure, calculate the maximum wind speed from equation (10.5), use that to make a new estimate of the surface pressure at the location of maximum winds, use that

to make a new estimate of E, plug that back into (10.5), and keep going around this loop. Eventually, the estimates of wind speed and surface pressure converge to the correct answer, unless the sea surface is extremely hot and/or T_{cold} is very small. In that case, there is no solution to the equation. What happens under these conditions is a runaway feedback: the lower the surface pressure, the more heat input results from the isothermal expansion of inflowing air; this additional heat input makes the storm more intense, dropping the surface pressure even lower, resulting in yet more heat input from isothermal expansion, and so on. The additional dissipation of kinetic energy simply cannot keep up. Although the conditions under which this would happen are far from those found on earth, we have simulated them with a computer model. The resulting storms, which we nicknamed *hypercanes*, collapse down to very tight vortices, with eyes only a few kilometers in diameter and with wind speeds approaching or even exceeding the speed of sound. These storms, rather than extending upward only into the very low stratosphere, would reach up to 30–40 km altitude, where they would deposit vast quantities of water in a layer of air that is normally very dry. This could cause a series of chemical reactions that would largely destroy the ozone layer that protects us from lethal ultraviolet radiation.

Besides E, we need to know the sea surface temperature, T_{hot}, and the temperature where air is flowing out of the top of the storm, T_{cold}. To estimate these, we need to know the vertical temperature profile of the atmosphere in the distant environment of the storm. Fortunately, this can be provided by weather balloons and other means.

Finally, we need to know the ratio C_K/C_D. We know that at very low wind speeds, this

Figure 10.3: Map showing the maximum wind speed (in mph) achievable by hurricanes over the course of an average year, according to Carnot's theory of heat engines.

▶

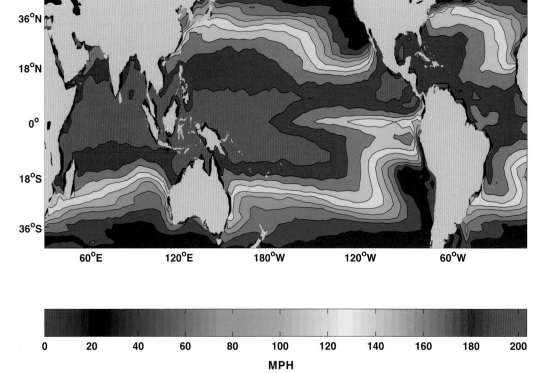

ratio is close to one. However, as the wind increases, waves develop on the ocean, making it rougher and increasing the drag coefficient, C_D. Up to wind speeds of around 60 mph, measurements show that this increase in C_D is enough to lower the ratio C_K/C_D to around 1/2. But using a value of 1/2 in equation (10.5) gives wind speeds that are too small to explain real hurricanes. Computer models show that at hurricane-force wind speeds, the ratio of exchange coefficients should be close to one, suggesting that some other processes are at work. Many research scientists believe that the missing ingredient is sea spray, which can transport enormous quantities of heat from the ocean to the atmosphere, as well as serving as an additional source of drag. The physics of air-sea interaction at hurricane wind speeds is an important research issue today.

Using climatological records of atmospheric temperature and humidity, and sea sur-

face temperature, and assuming some value of C_K/C_D, we can estimate the maximum wind speed that can be achieved in hurricanes. Figure 10.3 shows the maximum wind speed that can occur over the course of a year, assuming monthly mean climatological conditions. Deep in the Tropics, hurricanes can apparently have winds as high as 200 mph. The potential intensity of storms decreases rapidly, poleward of around 30° latitude. Comparing this map to the map of genesis locations (see Chapter 14), it is clear that hurricanes only develop where the potential intensity is large.

The hurricane Carnot heat engine, with its recycling of waste heat, is one of the most efficient natural generators of power on earth. The amount of power dissipated by a typical mature Atlantic hurricane, as given by equation (10.2) integrated over the whole surface area covered by the storm, is on the order of 3 trillion Watts, which happens to be very nearly equal to the worldwide electrical generation capacity as of January 1996. This is enough to light 30 billion 100-Watt light bulbs. A Pacific supertyphoon can dissipate ten times this power.

As Carnot states at the opening of this chapter, the flow of heat drives almost everything that happens in the earth's atmosphere and oceans, and also deep within its interior. Among these phenomena, an intense hurricane comes closest to Carnot's ideal heat engine.

▲

HMS *Egmont* dismasted in
the hurricane of October 6,
1780, near St. Lucia.
Mezzotint, drawn by
Lt. William Elliot, 1784.

I t began about dusk, at North, and raged very violently till ten o'clock. Then ensued a sudden and unexpected interval, which lasted about an hour. Meanwhile the wind was shifting round to the South West point, from whence it returned with redoubled fury and continued so till near three o'clock in the morning. Good God! what horror and destruction—it's impossible for me to describe—or you to form any idea of it. It seemed as if a total dissolution of nature was taking place. The roaring of the sea and wind—fiery meteors flying about in the air—the prodigious glare of almost perpetual lightning—the crash of the falling houses—and the ear-piercing shrieks of the distressed, were sufficient to strike astonishment into Angels. A great part of the buildings throughout the Island are levelled to the ground—almost all the rest very much shattered—several persons killed and numbers utterly ruined—whole families running about the streets unknowing where to find a place of shelter—the sick exposed to the keeness of water and air—without a bed to lie upon—or a dry covering to their bodies—our harbour is entirely bare. In a word, misery in all its most hideous shapes spread over the whole face of the country.—A strong smell of gunpowder added somewhat to the terrors of the night; and it was observed that the rain was surprisingly salt. Indeed, the water is so brackish and full of sulphur that there is hardly any drinking it.

—Alexander Hamilton (1755–1804), writing to his father about the famous hurricane
of 1772. Hamilton was born at Nevis and spent his teenage years at Saint Croix.

11   The Hurricanes of 1780

*The midnight horrors of the scene were viewed as the last convulsions
of an expiring world.*

—The Reverend George Wilson Bridges, Jamaica, 1780

In October 1780, the Caribbean was raked by three violent hurricanes, the second of which ranks as the deadliest storm ever to affect the Western Hemisphere. These events greatly altered the political and economic history of the region and further weakened the British Navy at a time when it was engaged in the American Revolutionary War.

THE SAVANNA-LA-MAR HURRICANE

The first hurricane began somewhere in the far southern Caribbean on October 1 and moved northwestward, striking Jamaica on the afternoon of the third. Early in its life, it sunk the British transport ship *Monarch*, killing all of its crew as well as several hundred Spanish prisoners onboard. As the storm approached the Jamaican port city of Savanna-la-Mar, curious residents lined up to watch the building seas. With no warning, a storm surge estimated to have been 20 ft high swept across the coast, engulfing the onlookers and hurling large ships far inland. The storm laid waste to the city and surrounding sugar plantations. Few structures withstood the onslaught, and many people perished in collapsing buildings, including the city's courthouse. In the port village of Lucea, about 20 mi north of Savanna-la-Mar, 400 people died and all but two houses were demolished; even the trees and shrubs were swept away. Further north, at Montego Bay, another 360 were killed.

A graphic and moving account of this storm was written by the Anglican clergyman George Wilson Bridges:

> The sea seemed mingled with the clouds, while the heaving swell of the earth, as it rolled beneath
> its bed, bore the raging floods over their natural boundaries, overwhelmed the coasts, and retreating with irresistible force, bore all before them. To the distance of half a mile, the waves carried

Figure 11.1: Tracks of the three hurricanes of October 1780. Numbers indicate dates. These tracks are based on ship and land observations, from Tannehill (1952).

▶

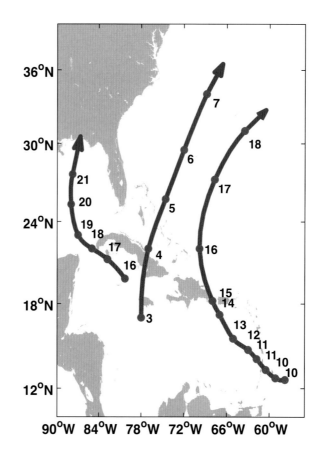

and fixed vessels of no ordinary size, leaving them the providential means of sheltering the house-less inhabitant. Not a tree, or bush, or cane was to be seen: universal desolation prevailed, and the wretched victims of violated nature, who would obtain no such shelter, and who had not time to fly to the protecting rocks, were either crushed beneath the falling ruins, or swept away, and never heard from more. The shattered remains of houses, whose tenants were dead or dying—the maddening search for wives and children, who were lost—the terrific howling of the frightened negroes, as it mingled with the whistling but subsiding winds—and the deluged state of the earth, strewed with the wreck of nature, and ploughed into deep ravines, was the scene which daylight ushered in; and, as if to mock the misery it had caused, the morning sun was again bright and cheerful.

As the storm approached Cuba, it wrecked the British frigate *Phoenix*, dashing it ashore at Cabo de la Cruz and killing 200 of its crew. The ships-of-the-line *Victor* and *Barbadoes*, which had sailed from Montego Bay on September 29, were never heard from again; nor was the *Scarborough*, which had left on October 1. The HMS *Ulysses*, also out of Montego Bay, was saved by her crew, who tossed all the guns off her upper deck, thus lowering her center of gravity. The *Stirling Castle* was also overtaken and crippled by the storm, but she struggled valiantly onward, only to be overtaken and sunk in the central North Atlantic by the second hurricane.

All told, some 3,000 lost their lives in this first storm, among them more than 1,500 sailors. The Savanna-la-Mar Hurricane remains one of the worst disasters in Jamaican history.

The Great Hurricane of 1780

The second of the three storms, known simply as the Great Hurricane of 1780, was the single deadliest storm ever to have affected the Western Hemisphere. Between October 10 and 18, this cataclysmic hurricane carved a path of destruction from Barbados to Bermuda. In Barbados, nearly every building on the island was leveled by the storm, and production of sugar and rum—vital to the local economy—was drastically curtailed, not recovering for another four years. The *Barbados Mercury* reported that "in most plantations all the buildings, the sugar mills excepted, are laid level with the earth, and that there is not a single estate on the island which has entirely escaped the violence of the tempest." More than 4,300 inhabitants died, and survivors were so traumatized that six months later a British newcomer reported, "The melancholy appearance of every Person & thing, struck me with a degree of Terror not easily to be described." Many settlers abandoned their plantations and returned to England, leaving the island's economy even further depressed.

On the smaller island of St. Vincent, to the west of Barbados, a storm surge estimated to have been 6 m (20 ft) high washed entire villages into the sea. To the north, the island of St. Lucia was flattened, killing some 6,000 inhabitants. More than 9,000 perished in Martinique, and the island's capital city of St. Pierre was almost completely demolished. An English fleet anchored off St. Lucia virtually disappeared in the tempest, while fifteen Dutch ships were lost to the south, in Grenada. The crew of a British vessel dashed onto Martinique were captured by French slaves but later released by a sympathetic governor, the Marquis de Bouillé.

The murderous storm continued inexorably northwestward, laying waste to the tiny island of St. Eustatius and killing between 4,000 and 5,000. After inflicting heavy damage and causalities in Puerto Rico and Dominique, the storm headed toward Bermuda, where it sunk or incapacitated several English ships.

The Great Hurricane of 1780 took more than 22,000 lives and struck a nearly fatal blow to the economy of the Caribbean. It so decimated the British fleet that the English presence in the western North Atlantic was thereafter significantly reduced. Although not meteorologically unusual, the hurricane's path took it directly over several of the most populous islands of the region. Not for more than two hundred years would the death toll of an Atlantic hurricane again exceed 10,000.

Solano's Hurricane

The third of the trio of intense hurricanes of October 1780 was first noted near Jamaica on the fifteenth. Progressing northwestward into the Gulf of Mexico, it dissipated over the southeastern United States around the twenty-second. It is most notable for having defeated a Spanish plan to take the Florida panhandle from the British. Under the command of Field Marshall Don Bernardo de Gálvez and Admiral Don José Solano, an armada of some 64 warships, transport ships, and supply vessels carrying some four thousand soldiers set out from Havana on the sixteenth with the objective of taking Pensacola. The slow-moving hurricane, known to history as "Solano's Hurricane," began to affect the fleet on the seventeenth and proceeded to wreak havoc over the succeeding five days, scattering and severely damaging the ships. Although nearly two thousand died, both Gálvez and Solano survived, returning finally to Havana. Gálvez returned to capture Pensacola the following May.

These three storms mark 1780 as in many ways the worst hurricane year in history. News of these tempests traveled far and wide and sealed the reputation of the Caribbean as a dangerous place for trade and habitation. The year was a turning point in Caribbean history, marking the end of a long period of prosperity and the beginning of an episode of economic and cultural decline.

▲

The Deluge. Oil painting by British artist Francis Danby, 1840.

S o when an angel by divine command
With rising tempests shakes a guilty land,
Such as of late o'er pale Britannia past,
Calm and serene he drives the furious blast;
And, pleas'd th' Almighty's orders to perform,
Rides in the whirlwind, and directs the storm.

—Joseph Addison (1672–1719),
    from "The Campaign, a Poem to His Grace
    the Duke of Marlborough"

A Battle Averted, Samoa, 1889

▲

The German warship Adler *overturned on a reef after the hurricane of March 1889, Apai Harbor, Samoa. Photograph ca. 1910–12. The* Calliope, *a British ship celebrated in a ballad (right), weathered the storm unharmed.*

Throughout the late nineteenth century, various Western powers contended for control of the South Pacific islands. One of the many island groups under dispute was Samoa. Between 1847 and 1861, Great Britain, the United States, and Germany all laid claims to the territory and established diplomatic missions. Part of their strategy involved siding with local tribal chiefs and arming them to battle rival chiefs supported by rival Western nations. Finally, in 1873, U.S. special agent Colonel A. B. Steinberger negotiated a peace, helped to draft a European constitution, and then ruled as a virtual dictator until his arrest and deportation by the British in 1875. The three Western powers remained in delicate balance until 1886, when Germany, with the consent of the British, landed naval forces in Western Samoa and attempted to establish control of the islands. The Germans were not, however, popular with the Samoans, who rebelled against what they perceived as heavy-handed rule. In 1888, the United States landed forces in Samoa, and all three Western powers dispatched naval forces to the islands. Matters drew to a head in November 1888, and for the next four months Germany and the United States were poised to go to war.

March 15, 1889, found many Western warships anchored in the exposed harbor of Apia. These included the U.S. Navy's Pacific Station flagship, USS *Trenton*, and the smaller U.S. warships *Vandalia* and *Nipsic*, present as a "show of force" in opposition to the German corvette *Olga* and gunboats *Adler* and *Eber*. Also anchored in Apia harbor were the British Royal Navy corvette *Calliope* and several civilian vessels.

Late on the fifteenth, the weather deteriorated rapidly, and although several naval officers were concerned about hurricanes, they were assured by natives that it was not hurricane season. Partly for this reason, and partly out of an unwillingness to show weakness to the rival naval forces, most of the ships remained at anchor. An exception was the comparatively modern British ship *Calliope*, which had just returned from exercises and had a full head of steam.

The tropical cyclone that struck Samoa on the night of the fifteenth and morning of the sixteenth wreaked havoc with the Western warships. With the exception of the *Calliope*, which managed with great difficulty to steam out of the harbor into the teeth of the storm, the ships proved unable to deal with the tempest. The *Eber*, the smallest of the warships, was blown into the reef and completely destroyed, with the loss of nearly all her crew. The *Adler* and *Vandalia* also suffered heavy personnel casualties and were wrecked beyond recovery. The *Trenton*, whose steam power plant was extinguished by water entering through her low hawse pipes, dragged her anchors and was also wrecked, but losses among her men were light. The *Olga* and *Nipsic* were run ashore, though both were later hauled off and repaired.

The storm killed about 50 U.S. sailors and 90 Germans, but ended any possibility of war in Samoa. A truce was negotiated, though the political climate remained unstable for many years afterward.

Ballad of the *Calliope*

By the far Samoan shore,
　　Where the league-long rollers pour
All the wash of the Pacific on the coral-guarded bay,
　　Riding lightly at their ease,
　　In the calm of tropic seas,
The three great nations' warships at their anchors proudly lay.

Riding lightly, head to wind,
　　With the coral reefs behind,
Three German and three Yankee ships were mirrored in the blue;
　　And on one ship unfurled
　　Was the flag that rules the world—
For on the old *Calliope* the flag of England flew.

　　When the gentle off-shore breeze,
　　That had scarcely stirred the trees,
Dropped down to utter stillness, and the glass began to fall,
　　Away across the main

Lowered the coming hurricane,
And far away to seaward hung the cloud-wrack like a pall.

If the word had passed around,
    "Let us move to safer ground;
Let us steam away to seaward"—then this tale were not to tell!
    But each Captain seemed to say
    "If the others stay, I stay!"
And they lingered at their moorings till the shades of evening fell.

    Then the cloud-wrack neared them fast,
    And there came a sudden blast,
And the hurricane came leaping down a thousand miles of main!
    Like a lion on its prey,
    Leapt the storm fiend on the bay,
And the vessels shook and shivered as their cables felt the strain.

    As the surging seas came by,
    That were running mountains high,
The vessels started dragging, drifting slowly to the lee;
    And the darkness of the night
    Hid the coral reefs from sight,
And the Captains dared not risk the chance to grope their way to sea.

    In the dark they dared not shift!
    They were forced to wait and drift;
All hands stood by uncertain would the anchors hold or no.
    But the men on deck could see,
    If a chance for them might be,
There was little chance of safety for the men who were below.

    Through that long, long night of dread,
    While the storm raged overhead,
They were waiting by their engines, with the furnace fires aroar;
    So they waited, staunch and true,
    Though they knew, and well they knew,
They must drown like rats imprisoned if the vessel touched the shore.

    When the grey dawn broke at last,
    And the long, long night was past,
While the hurricane redoubled, lest its prey should steal away,
    On the rocks, all smashed and strown,
    Were the German vessels thrown,
While the Yankees, swamped and helpless, drifted shorewards down the bay.

Then at last spoke Captain Kane,

"All our anchors are in vain,

And the Germans and the Yankees they have drifted to the lee!

Cut the cables at the bow!

We must trust the engines now!

Give her steam, and let her have it, lads! we'll fight her out to sea!"

And the answer came with cheers

From the stalwart engineers,

From the grim and grimy firemen at the furnaces below;

And above the sullen roar

Of the breakers on the shore

Came the throbbing of the engines as they laboured to and fro.

If the strain should find a flaw,

Should a bolt or rivet draw,

Then-God help them! for the vessel were a plaything in the tide!

With a face of honest cheer

Quoth an English engineer,

"I will answer for the engines that were built on old Thames-side!

"For the stays and stanchions taut,

For the rivets truly wrought,

For the valves that fit their faces as a glove should fit the hand.

Give her every ounce of power;

If we make a knot an hour

Then it's way enough to steer her, and we'll drive her from the land."

Like a foam-flake tossed and thrown,

She could barely hold her own,

While the other ships all helplessly were drifting to the lee.

Through the smother and the rout

The *Calliope* steamed out

And they cheered her from the *Trenton* that was foundering in the sea.

Ay! drifting shoreward there,

All helpless as they were,

Their vessel hurled upon the reefs as weed ashore is hurled,

Without a thought of fear

The Yankees raised a cheer—

A cheer that English-speaking folk should echo round the world.

—Andrew Barton "Banjo" Paterson (1864–1941)

12　　Hurricane Intensity

Blow, winds, and crack your cheeks! rage! blow!
You cataracts and hurricanoes, spout
Till you have drench'd our steeples, drown'd the cocks!
You sulphurous and thought-executing fires,
Vaunt-couriers to oak-cleaving thunderbolts,
Singe my white head! And thou all-shaking thunder
Smite flat the thick rotundity o' the world!

　　　—Shakespeare, *King Lear*

The Carnot heat engine theory (Chapter 10) predicts the maximum wind speed achievable by hurricanes. But how accurate is this prediction? What controls the intensity of real storms?

Very few hurricanes reach the theoretical maximum wind speed. There are many reasons for this, not all of them well understood. One simple factor is time. In nature and in computer simulations, it typically takes five days to a week for a tropical depression to strengthen into a full-blown hurricane. Many factors can disrupt it before it has a chance to realize its potential. For example, it can pass over land or move over colder water, where the potential intensity is small or zero, or it can encounter unfavorable atmospheric conditions that cause it to stop intensifying or to weaken.

Thanks to hurricane hunter aircraft missions, there are reliable measurements of the peak wind speeds achieved in Atlantic tropical cyclones for the last 40 years or so, and over the western North Pacific between 1965 and 1987. We can compare the maximum wind speed recorded during the lifetime of each storm with the Carnot potential intensity calculated from atmospheric and oceanic data at the place and time each storm reached its peak intensity. To make the comparison meaningful, we exclude all storms that made landfall while they were still intensifying and storms that moved rapidly over colder water.

We calculate the ratio of each storm's peak wind speed to its theoretical maximum wind speed. Figure 12.1 shows the number of events whose ratio exceeds the value on the bottom axis. For example, there were 230 events whose actual maximum wind speed exceeded 30 percent of their potential wind speed, and only about 10 events whose actual wind speed exceeded 90 percent of their potential wind speed.

The distribution tends to fall along two straight lines, breaking at about 45 percent. No

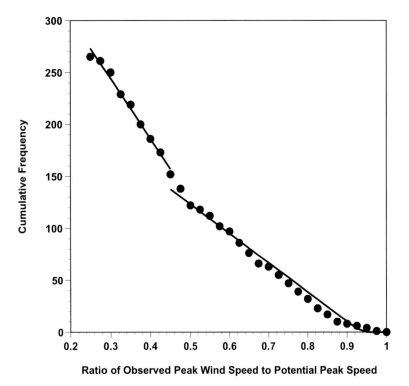

Figure 12.1: Total number of events in the Atlantic, from 1958 to 1997, and the western North Pacific, from 1970 to 1987, whose ratio of actual to potential maximum wind speed exceeds the value on the x axis.

▶

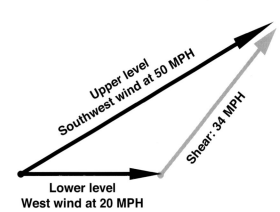

Figure 12.2: This diagram shows how shear is defined. Here a westerly wind of 20 mph is blowing near the surface, while higher up the wind is from the west-southwest at 50 mph. The shear between these two levels is a vector pointing toward the northeast with a magnitude of 34 mph.

▶

one knows why this break occurs, but it is curious that it nearly corresponds with the transition from tropical storm to hurricane strength.

As might be expected, no storms exceed their theoretical speed limit. But it is equally clear that the vast majority of storms never get close to their potential, even when one excludes storms whose intensity is limited by landfall or travel over cold water. Something else is holding them back. Meteorologists suspect two main culprits:

VERTICAL WIND SHEAR

Forecasters have long known that vertical shear of the horizontal wind is a major impediment to tropical cyclone formation and intensification. Shear is defined as the magnitude of the vector

Figure 12.3: Tropical Storm Barry in the central Gulf of Mexico on August 2, 2001. Note that the deep convective clouds are arranged in an open arc around the storm center rather than completely enclosing it. This shows the effect of vertical wind shear, which, at this time, was directed toward the north.

▶

difference between the wind at two different altitudes. To visualize this, just draw two arrows aligned with the wind at each altitude and having lengths proportional to the wind speeds at the respective altitudes. Now draw an arrow connecting the tips of the first two arrows. The length of this arrow is proportional to the wind shear. In the example shown in Figure 12.2, a west wind of 9 m/s (20 mph) is blowing near the surface, while high up in the atmosphere, the winds are from the west-southwest (actually, from 240 degrees on the compass) at 22 m/s (50 mph). The strength of the shear is, in this case, 15 m/s (34 mph).

It is not clear what measure of shear is best to use when trying to predict hurricane intensity. In practice, the shear between altitudes of about 1.5 km (5,000 ft) and 10 km (33,000 ft) is often used, but it may be that the shear between other altitudes is important. The appropriate measures of shear might be more complex, involving differences among three or more levels. Uncertainty

about the best measure of shear reflects our collective ignorance about the precise effects of shear on hurricanes.

Shear is, however, known to make the storm circulation lose its circular symmetry. In general, the strongest updrafts will be on the *downshear* side of the eye; i.e., the side to which the shear vector is pointing. (In the example of Figure 12.2, this would be the northeast side of the eye.) Convection may be weak or even absent on the upshear side of the eye. Figure 12.3 shows a satellite image of a hurricane affected by shear from the south (i.e. from the bottom of the image). The circulation center is just south of the thunderheads visible near the center of the image.

Computer models of hurricanes clearly show this effect. Figure 12.4 shows vertical slices through two computer simulations of hurricanes; in the first, the vertical shear is weak while in the second it is strong. In the weak shear case (left

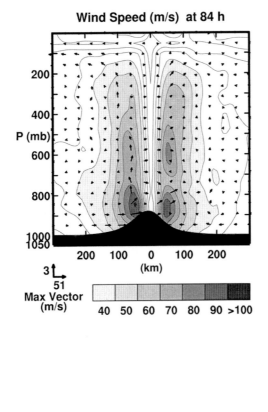

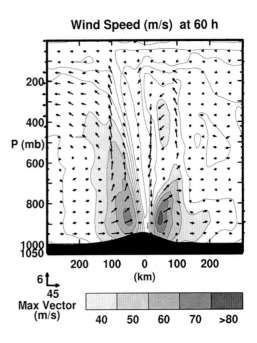

Figure 12.4: Cross sections through a computer simulation of a hurricane. In each panel, the arrows show the vertical and radial wind, and the shading shows the strength of the wind blowing around the storm's axis. The diagrams extend from the center outward to 300 km (185 mi) and from the sea surface upward to about 18 km (60,000 ft); the vertical axis is actually labeled with pressure, which decreases upward. The left panel shows a simulation with weak wind shear, while the right panel shows one with strong shear directed from right to left.

►

panel), the wind and updraft are distributed nearly symmetrically around the center, while in the strong shear case (right panel), most of the upward motion is on the downshear (left) side of the eye, although, curiously, the winds are slightly stronger on the upshear side.

Besides disrupting its circular symmetry, shear has other potentially detrimental effects on the storm. The erosion of the high, dense overcast over some or all of the eyewall region allows infrared radiation to escape to space, cooling the core of the storm and thereby weakening it. Shear can also force dry, low energy air from the storm's environment to swirl into its core at middle levels, reducing the core entropy and weakening the storm. From the Carnot cycle point of view,

the second leg of the cycle is disrupted, so that some of the very warm, moist air ascending in the eyewall mixes out of the core at middle levels rather than ascending all the way to the storm top. Thus the effective cold reservoir temperature reflects the middle-level ambient temperature—around 270 Kelvin (27°F)—rather than the 200 Kelvin (-100°F) typical of the storm-top environment. This can profoundly reduce the strength of the winds. The injection of dry air into the storm core is sometimes referred to as *ventilation.*

We are just beginning to understand how wind shear adversely affects hurricanes. Even if we understood hurricanes better, it is not certain that we could make better forecasts of hurricane

Figure 12.5: A satellite infrared image showing the sea surface temperature distribution of the western North Atlantic on September 2 and 3, 1996, shortly after Hurricane Edouard passed along the black curve shown in the image. The color scale is shown at lower right. (The white patches are clouds.) Along and to the right of the path of Edouard, the sea surface temperature has decreased by as much as 5°C.

▶

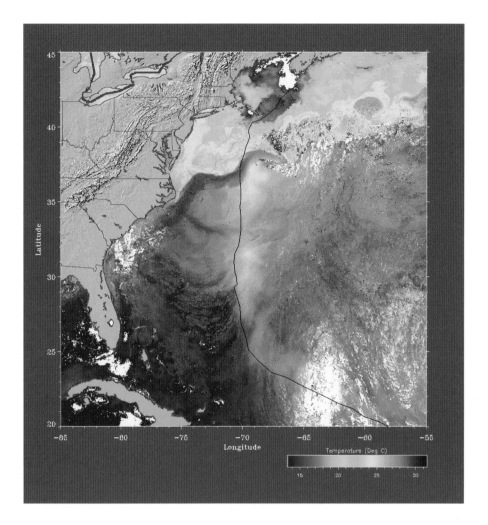

intensity, because it is difficult to observe and forecast winds over the tropical oceans where measurements are sparse. In recent years, progress has been made using satellites to track clouds and other features, from which the speed and direction of the wind can be deduced.

OCEAN INTERACTION

The Carnot theory assumes that the sea surface temperature underneath hurricanes does not change as they pass over. Yet the ocean is often observed to cool, especially to the right of the storm track (to the left in the Southern Hemisphere). Figure 12.5 shows the sea surface temperature just after the passage of Hurricane Edouard of 1996, whose track is depicted by the thin black line.

It is tempting to conjecture that this ocean cooling is caused by the loss of heat to the atmosphere, which, after all, powers the storm. But calculations show that such heat loss would, on average, cool the upper layers of the ocean by only about 0.1°C (0.2°F); not nearly as much as has been observed. The ocean is an enormous heat reservoir, and even a hurricane does not deplete it very much.

The main cause of the cooling is the mixing to the surface of cold water from deeper in the ocean. In the Tropics, the warmest water is contained in a thin layer at the surface of the

Figure 12.6: Wind and current directions at two points (A and B) to the left and right of the track of a hurricane in the Northern Hemisphere. Read figure from right to left, as the hurricane passes in between points A and B. The solid black arrows give the wind direction, and the white arrows show the ocean current direction.

▶

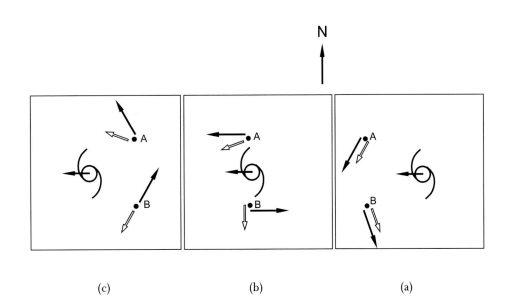

(c)   (b)   (a)

ocean; this is called the *mixed layer* because many water properties, including temperature, are uniformly mixed within this layer. The layer is kept continually mixed by turbulence caused by wind and breaking waves. The thickness of the mixed layer varies, from less than 15 m in parts of the Gulf of Mexico to over 100 m in the western Caribbean and the western North Pacific. Below the mixed layer, temperature drops off rapidly with depth, decreasing on average about 8°C (14°F) each 100 m (330 ft).

Powerful storms affect the upper ocean in several ways. First, breaking waves add more turbulence to the mixed layer; this turbulence eats away at the mixed layer base, stirring in colder water from below. More importantly, the strong winds blowing across the sea surface drive ocean mixed layer currents that may reach speeds of 2 m/s (5 mph). The large vertical shear of the ocean current velocity across the base of the mixed layer causes turbulence there, which also stirs in colder water from below. Research shows that this second mechanism is the main cause of the sea surface cooling.

The shift of the cooling toward the right side of the storm track (in the Northern Hemisphere) is due to the rotation of the earth. This can be understood as follows.

All moving objects not subjected to a force move in straight lines with respect to an observer fixed in space. But an observer on the earth is not fixed in space since the earth is rotating. An object moving in a straight line as seen by a hypothetical observer fixed in space will appear to be curving to an observer on the rotating earth. Because we observers are stationed on the earth, we find it more convenient to imagine that the object experiences a deflecting force. This invented force is called the Coriolis force, after the nineteenth-century French scientist who first formulated it. It acts perpendicular to the motion of any object, to the right of its motion in the Northern Hemisphere.

When the wind pushes on the sea surface, the water at first responds by moving in the direction of the wind. But with time, the Coriolis force turns the water to the right. If the wind friction lasted for only a short time, we would observe a

Figure 12.7: Horizontal map
showing the ocean currents
and mixed layer depths under
a hurricane moving along the
black line from right to left,
generated from a computer
simulation. The storm is
currently located at the black
asterisk. The ocean current
direction is given by the black
arrows, whose length is proportional to its speed. The
colors show the thickness of the
ocean mixed layer (in meters),
with color scale at bottom.

▶

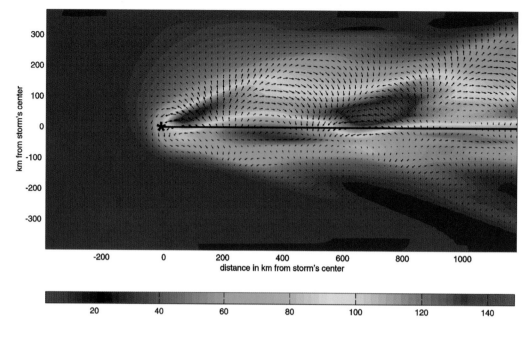

current that first moves in the same direction as the wind but gradually turns to the right. Given enough time, the current would go all the way around the compass until it was headed in the same direction it started. The time it would take to do this is the same time it takes an ideal pendulum to make one complete turn, and is therefore called a *pendulum day*. At 30° latitude, a pendulum day is 24 hours; at the poles it lasts only 12 hours, and at the equator there is no Coriolis force, so the pendulum day lasts forever.

Consider a patch of seawater located at point A to the right of the path of an oncoming hurricane moving from east to west, as shown in Figure 12.6. The ocean at point A first experiences a wind from the northeast and so begins to flow toward the southwest (panel *a*). With time, the Coriolis force turns the current so that it is headed more toward the west (panel *b*). But at the same time, the hurricane is passing by to the south, so the wind has veered around to the east. The wind keeps pushing the water in the direction it is already headed, so it flows faster. As the

storm passes by to the west (panel *c*), the winds switch around to blow from the southeast, and the Coriolis force has turned the ocean current so that it, too, is headed toward the northwest. Once again, the wind is pushing water in the direction it is already heading, so it flows ever faster.

To the left of the storm track (point B), the reverse is true. The initial northwest wind pushes water toward the southeast, but the Coriolis force gradually turns the current toward the south. At the same time, the hurricane passes by to the north and produces a westerly wind, which is neither aiding nor opposing the current. Sometime later, the Coriolis force has rotated the current toward the southwest at a time when the hurricane winds are blowing from the southwest, opposing the current, which therefore slows down.

Thus to the right of the storm track, ocean currents are continuously accelerated by the wind, but left of the track, the wind and the currents are often opposed and the currents are therefore weaker. The weaker the current, the

Figure 12.8: Evolution of the maximum wind speed in Tropical Storm Barry of 2001. The blue curve shows the wind speed estimated from observations, and the other curves show various attempts to simulate this with a computer model, including a simulation without ocean mixing (green), a simulation with ocean mixing but without wind shear (red), and a simulation with both ocean mixing and shear (light blue).

▶

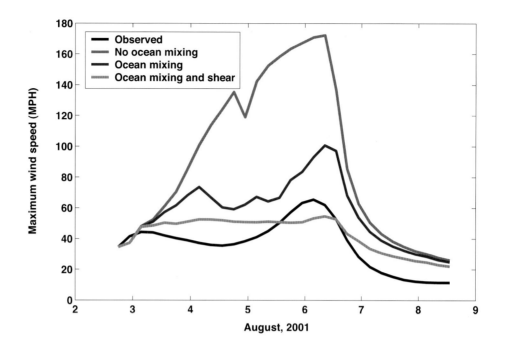

less turbulence is generated at the base of the ocean mixed layer, and the less cold water is stirred into it.

Figure 12.7 shows the response of the upper ocean to a hurricane moving from east to west, from a computer simulation using a coupled ocean-atmosphere hurricane model. The colors show the local depth of the mixed layer, and the black arrows show the direction and speed of the ocean current. The mixed layer deepens greatly, particularly to the right of the track, as cold water is turbulently mixed into its base. The strongest currents are also to the right of the track, and they can be seen turning around the compass as the storm passes by. Although the water temperature is not shown, the coldest water is where the mixed layer is deepest.

Although the coldest seawater is behind and to the right of the storm, the ocean under the eyewall is a few degrees Celsius colder than the undisturbed water ahead of the storm. This ocean cooling reduces the amount of energy that can be added to the inflowing air by evaporation.

In fact, a mere 2.5°C drop in ocean temperature under the eyewall suffices to stop all energy input to the storm. Therefore even a 1°C ocean temperature decrease under the eyewall will noticeably decrease the storm's intensity.

Experiments with computer models that allow for the cooling of the ocean show that, all other things being equal, hurricanes will be less intense if they are moving slowly, they have a large eye, and/or the ocean mixed layer ahead of the storm is thin. Also, the more intense the hurricane would otherwise be, the greater the relative effect of ocean mixing: intense storms churn up more cold water than weak ones.

Computer models demonstrate that both vertical wind shear and ocean mixing strongly affect hurricane intensity. Figure 12.8 shows the time evolution of the peak wind speed in Tropical Storm Barry of 2001, as simulated by a simple coupled model of the atmosphere and ocean. (Barry is the same storm shown in the satellite image in Figure 12.3; it is clearly being affected by shear.) In this model, shear is assumed to inject dry air into the core of the storm.

Tropical Storm Barry never achieved hurricane strength, topping out at about 27 m/s (60 mph) just before making landfall in the Florida panhandle. With neither shear nor ocean cooling, the simulated storm becomes too intense, rivaling Camille of 1969. Including ocean cooling in the model brings the peak intensity down to about 45 m/s (100 mph), still much greater than observed. When both wind shear and ocean cooling are accounted for, the simulated intensity, while not perfect, is much better than simulations omitting one or both effects.

There are some subtleties in accounting for ocean mixing. If the ocean is not very deep near the coast, the mixed layer may extend right to the bottom for some distance offshore. When this happens, a hurricane approaching the coast may suddenly encounter warm water with no cold water beneath it. With no mixing of cold water from the depths, the sea surface remains warm and the hurricane may intensify as it approaches land.

An important practical problem in accounting for ocean cooling is the paucity of ocean temperature measurements below the surface. The ocean, unlike the atmosphere, strongly absorbs radio waves, making it nearly impossible to transmit data from subsurface sensors. For the same reason, satellites do not "see" more than a few meters below the surface. Taking and recording measurements below the surface is a difficult, expensive, and time-consuming process. Weather balloons, aircraft, and satellites regularly and routinely measure the properties of the atmosphere many times each day, but in the ocean, the depths must be probed by instruments deployed from ships, which are expensive to operate, or by

buoys deployed from special airplanes, which are likewise expensive. Consequently, the interior of the ocean is probed very rarely and irregularly. Coupled hurricane models like the one used to produce Figure 12.8 have to assume average ocean conditions for the location and time of year in question.

The upper ocean temperature does change, however. Strong, narrow currents like the Gulf Stream meander with time, and the meanders can become so pronounced that they pinch off into isolated eddies, either cold or warm, that can then migrate elsewhere. In the Tropics, these eddies may be very difficult to detect from satellite because they hardly affect surface temperature: they are just places where the mixed layer is deeper or shallower than normal. A hurricane passing over a warm eddy produces much less surface cooling, because the warm water there is much deeper than normal; this can lead to unexpected intensification of the storm.

An important feature of the Gulf of Mexico is the Loop Current. This current enters the Gulf through the Straits of Yucatán and may take a quick right turn around Cuba and exit through the Straits of Florida, where it becomes the Gulf Stream. At other times, it continues some distance up into the Gulf, making a hairpin turn to the right and flowing parallel to the west coast of Florida before exiting through the Straits of Florida. A warm eddy sometimes develops at the hairpin turn, breaking off from the main current and drifting slowly west or southwest. All these changes take months. But lack of good measurements makes it difficult to locate the currents and eddies at any given time.

An intriguing case is that of Hurricane Camille of 1969 (see Chapter 26). Camille moved north-northwestward from the Yucatán Straits across the Gulf to make landfall in Mississippi as the most intense hurricane in history to strike the U.S. mainland, packing wind speeds of almost 90 m/s (200 mph). The evolution of its wind speed was quite similar to that of the "no ocean mixing" simulation of Tropical Storm Barry (Figure 12.8). Normally, a hurricane moving at Camille's speed over the Gulf should have mixed a great deal of cold water to the surface, limiting its intensity to around 45 m/s (100 mph). It is possible that Camille, by an unhappy accident, traveled right up the axis of the Loop Current, whose warm water runs so deep that little or no cold water can be mixed to the surface by storms.

Forecasters have long struggled without much success to forecast hurricane intensity change. The skill of hurricane intensity forecasts, measured against control forecasts that are based on statistics of previous storms, has not improved nearly as much as the skill of hurricane track forecasts. This may be partly because of the difficulty of measuring and forecasting vertical shear in the vicinity of the storm, and partly because of inadequate knowledge of the temperature of the upper ocean along the storm's track.

Several recent developments raise hopes for better hurricane intensity forecasts. One of these is the deployment of satellite-borne sea surface altimeters, remarkable instruments that measure the level of the sea surface to a precision of a few inches. Subtle variations of the sea surface height give clues about the underlying thermal structure of the upper ocean. New technology like this and better understanding of the intricate working of hurricanes may someday allow accurate forecasts of hurricane intensity one to three days in advance.

WRECK OF THE STEAMSHIP CENTRAL AMERICA.

APPALLING DISASTER.

On Saturday, September 12th, 1857, Capt. Herndon, bound to New York, from California, with the Pacific Mails,
Passengers and Crew, to the number of 592 persons, and treasure to the amount of over
$2,000,000, foundered in a hurricane, off Cape Hatteras.

Whole number on board, 592.   Number saved, 166.   Number on board whose names are known, 184.   Names unknown, 292.

▲

Wreck of the steamship
Central America,
September 12, 1857.
The ship foundered in a
hurricane off Cape
Hatteras carrying 626
passengers and 21 tons
of gold from the California
gold rush. Only about 60
people were rescued.
The gold was recovered
in 1989 in water over two
miles deep. Color
lithograph by J. Childs.

I n the dread Ocean, undulating wide,
Beneath the radiant Line that girts the Globe,
The circling Typhon, whirl'd from Point to Point,
Exhausting all the Rage of all the Sky,
And dire Ecnephia reign. Amid the Heavens,
Falsely serene, deep in a cloudy Speck
Compress'd, the mighty Tempest brooding dwells.
Of no Regard, save to the skilful Eye,
Fiery and foul, the small Prognostic hangs
Aloft, or on the Promontory's Brow
Musters its Force. A faint deceitful Calm,
A fluttering Gale, the Demon sends before,
To tempt the spreading Sail. Then down at once,

Precipitant, descends a mingled Mass
Of roaring Winds, and Flame, and rushing Floods.
In wild Amazement fix'd the Sailor stands.
Art is too slow. By rapid Fate oppress'd,
His broad-wing'd Vessel drinks the whelming Tide,
Hid in the Bosom of the black Abyss.
With such mad Seas the daring GAMA fought,
For many a Day, and many a dreadful Night,
Incessant, lab'ring round the stormy Cape;
By bold Ambition led, and bolder Thirst Of Gold.

—James Thomson (1700–1748),
    from "The Seasons, Summer"

13   Galveston, 1900

The opinion held by some who are unacquainted with the actual conditions of things, that Galveston will at some time be seriously damaged by some such disturbance, is simply an absurd delusion.

—Isaac Cline, Local Forecast Official and Section Director, U.S. Weather Bureau, Galveston, Texas, in the *Galveston News,* 1891

The storm will probably continue slowly northward and its effects will be felt as far as the lower portion of the middle Atlantic coast by Friday night.

—Forecast issued by U.S. Weather Bureau, 8:00 A.M., Thursday, September 6, 1900

Sunday, September 9, 1900, revealed one of the most horrible sights that ever a civilized people looked upon.

—Isaac Cline

The Galveston Hurricane of 1900 was by far the worst natural calamity in U.S. history. The city itself was almost completely destroyed, and the death toll of between 8,000 and 12,000 exceeds that of the 1906 San Francisco Earthquake, the 1889 Johnstown Flood, and the 1928 Okeechobee Hurricane combined. Between the badly bungled forecast of September 6 and the tragedy at Galveston two days later lies a tale of individual courage and misjudgment, of bureaucratic envy and xenophobia.

In 1900, Galveston was the premier city of Texas, the "New York of the Gulf," as the *New York Herald* proclaimed. It was competing with Houston to become the dominant city of the region, on a par with New Orleans. It exuded an air of terrific optimism and boasted an annual population growth of almost 3 percent. The city was built on Galveston Island, a narrow, low sandbar separating the Gulf of Mexico from Galveston Bay, its developers having ignored stories that an 1841 storm had submerged the entire island to such a depth that ships could cross it.

Sparse ship reports suggest that the storm that was to destroy Galveston formed in the central North Atlantic, about four hundred miles west of the Cape Verde Islands, on or about August 27. A ship encountering the storm the next day recorded winds of about 14 m/s (30 mph). By Thursday the thirtieth, the tropical storm was near Antigua in the Leeward Islands, where the observing station at St. Johns recorded a minimum pressure of 1010 mb (29.83"). Proceeding westward on September 1–4, the storm grazed the south coasts of Hispaniola and Cuba, inundating Santiago, Cuba, with 87 cm (24.34") of rain in just two days. Many miles of railway bedding were washed away in Jamaica as the storm passed by to its north.

The forecasters at Havana's Belen Observatory monitored the storm with interest. Father Gangoite, the observatory's director, issued a statement on September 1 expressing his view that the storm, while small, was of a type known to produce heavy rain in Cuba and to intensify rapidly after passing into the Florida Straits. The tropical storm crossed over western Cuba during the fourth, and the next day the Belen observers reported it near Havana, moving northwestward into the Gulf of Mexico.

It was during this time that politics threw a nasty wrench into the already shaky machinery of turn-of-the-century hurricane forecasting. Officials of the Washington D.C.–based U.S. Weather Bureau Central Office, headed by Willis Moore, had grown increasingly exasperated with Cuban forecasters, whom they regarded as inferior and alarmist. In reality, the Cubans had achieved a well-deserved reputation for skill in hurricane observation and forecasting, and the Weather Bureau sought to curtail the competition. As fate would have it, the conflict climaxed in late August, when Moore instituted a ban on all transmission of West Indian storm reports from the Bureau's own Havana office to its New Orleans office. On August 28, Moore wrote the Western Union in an attempt to get them to enforce the cable ban: "The United States Weather Bureau in Cuba has been greatly annoyed by independent observatories securing a few scattered reports and then attempting to make weather predictions and issue hurricane warnings to the detriment of commerce and the embarrassment of the Government service. I have reason to believe that they are copying, or contemplate doing so, data from our daily weather maps in New Orleans and cabling the same to Havana."

This letter exposes two of the fears that have compromised the objectivity of forecasts issued by government weather services from time to time since their inception: that a forecast of bad weather will have adverse economic consequences, and that the government service (i.e. its administrators) will be "embarrassed." To this must be added the xenophobia for which the Weather Bureau was notorious through the first century of its existence. (In the 1930s, the Bureau issued an internal memorandum barring from its local offices one Carl Gustav Rossby, probably the most famous meteorologist of his generation, on the grounds that he advocated drawing fronts on weather maps, a practice invented by the Norwegians and thus subject to the "not-invented-here" mentality of the Bureau.)

The final sentence in Moore's letter reflects an attitude toward publicly financed data that was to vanish early in the twentieth century, only to reemerge in Europe in the 1980s. This attitude holds that data collected by a government agency is owned by the government or its agency and should not be shared with other peoples or nations. By midcentury, virtually all of the world's governments had come to see the many benefits that accrue from treating weather data as a public good: by making such data easily and cheaply available, most countries get back far more than they give. This practice was so widespread and so beneficial that many countries traded weather data even when they were at war with each other. In light of this policy, Moore's sentiment seems particularly obtuse.[8]

Whatever its cause, the animosity of the U.S. Weather Bureau toward its Cuban colleagues no doubt contributed to serious forecast blunders made in the succeeding days. The track of the storm through late on September 5 showed it gradually turning north (Figure 13.1). Once they begin to recurve, most hurricanes continue to do so. Thus Weather Bureau forecasters were very much within

---

[8] It was fear of competition from the private sector rather than from other nations that led several European governments to restrict the availability of weather data beginning in the 1980s. Unfortunately, this policy continues at the time of this writing.

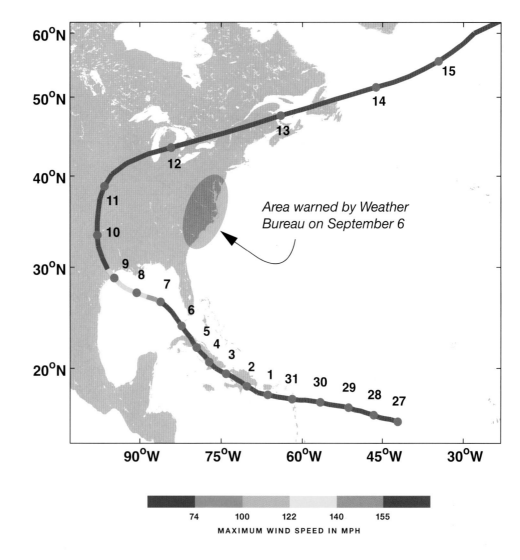

Figure 13.1: Track of the deadly Galveston Hurricane of August–September 1900. The red dots show the storm's position each day, beginning on August 27. The orange patch shows the area warned by the U.S. Weather Bureau on September 6, when the storm was just off Key West.

▶

Area warned by Weather Bureau on September 6

MAXIMUM WIND SPEED IN MPH

their rights to predict on September 5 that the storm would move northward and pose a threat to Florida and the East Coast of the United States. They shrugged off the Cuban forecasters' suggestion that the storm would move northwestward into the Gulf. So wedded were they to their prediction of continued recurvature that they missed what to an unbiased observer would have been an obvious clue. Late on the evening of Wednesday, September 5, the wind at Key West, Florida, which had been blowing a true gale from the northeast, suddenly weakened, and then shifted to the south while strengthening again. This was a sure sign that the storm center had passed close by to the southwest and then headed northwestward. Moreover, the strength of the wind at Key West and elsewhere in Florida showed that the storm had not been destroyed by its passage over Cuba. Even so, the next morning the Weather Bureau boldly asserted that the storm was 150 mi northeast of Key West, in complete contradiction to these observations. As late as that Thursday afternoon, the Bureau was warning fishermen in New Jersey to stay in port.

Out over the eastern Gulf, the storm was intensifying rapidly and heading toward Texas. The steamship *Louisiana*, out of New Orleans, ran straight into the storm center and at 1 P.M. Thursday, while the Bureau was hoisting warnings in New Jersey, recorded a minimum pressure of 973 mb (28.74"). Its captain estimated the wind speed at well over 45 m/s (100 mph). Late the next day, the steamship *Pensacola*, out of her home port in Florida and bound for Galveston, was savaged by the storm and experienced a minimum pressure of 966 mb (28.53"), though there is little indication that she passed through the center. The crew and passengers of these ships were the only mortals that knew the true intensity of the hurricane at this time. Ships were not yet equipped with the "wireless," which had been invented only a few years earlier. Fortunately, both vessels survived, and their weather reports were used to help reconstruct the meteorological history of the notorious storm.

Although the ship's reports could not reach the Weather Bureau until after they had docked, the Bureau could no longer ignore obvious indications that a tropical cyclone was moving westward or northwestward through the Gulf. For one thing, the storm had failed to materialize in Florida and along the East Coast, where the Bureau's forecasters had predicted it to. In addition, wind reports from observing stations along the Gulf coast clearly showed that a cyclonic circulation was present in the Gulf. Finally, at 9:35 A.M. Galveston time on Thursday, September 6, Willis Moore telegraphed the Bureau's offices along the Gulf coast to hoist storm warnings. The word "hurricane" did not appear in any of these transmissions, conforming to the Bureau's policy at that time, molded as it was by fear of diminishing commerce and, perhaps, in reaction to what was viewed as the fear mongering of the Cuban forecasters. Moreover, this warning was issued as Moore himself was still advising East Coast shipping to stay in port; clearly the forecasters were still clinging to their earlier predictions in spite of obvious signs that these had been erroneous.

In Galveston, the chief of the local Weather Bureau office, Isaac M. Cline, woke late Friday night to the portentous sound of heavy breakers on the beach, several city blocks away. The next morning, September 8, he found that the time interval between breakers was unusually long, suggesting that a storm was churning somewhere out in the Gulf. On the other hand, ambiguous transmissions from the Bureau's central office, while suggesting the possibility of the storm in the Gulf, also indicated that it was moving up the eastern seaboard. The brick-dust sky that Cline had been taught to believe presaged hurricanes was entirely absent, as were the high cirrus that normally run out ahead of tropical cyclones. The barometer had only fallen about 3.5 mb (0.1") from the previous evening. Yet the sea had already risen far enough to flood the seaward end of the city streets, despite an offshore wind from the north. Cline was now sufficiently alarmed to telegraph Washington, "such high water with opposing winds never observed previously."

And yet Cline's office issued no hurricane warnings. It was the Weather Bureau's strictly enforced policy that no hurricane warnings would be issued except by its central office in Washington. Bureau employees who violated this policy risked losing their jobs. But contributing to the lack of warning were Isaac Cline's own preconceptions about Galveston's vulnerability to hurricanes. Several prominent scientists, among them the oceanographer Matthew Fontaine Maury, had published papers purporting to demonstrate that storm surges of the kind that swept over the Ganges delta in

Figure 13.2: Computer simu-
lation of the evolution of the
maximum wind speed (top)
and minimum surface pres-
sure (bottom) of the Galveston
Hurricane of 1900. The top
panel also shows an estimate
of the actual maximum wind
speed in the storm, using
what few observations were
available. Surface pressures
recorded on two ships and at
Galveston are shown in the
bottom panel.

▶

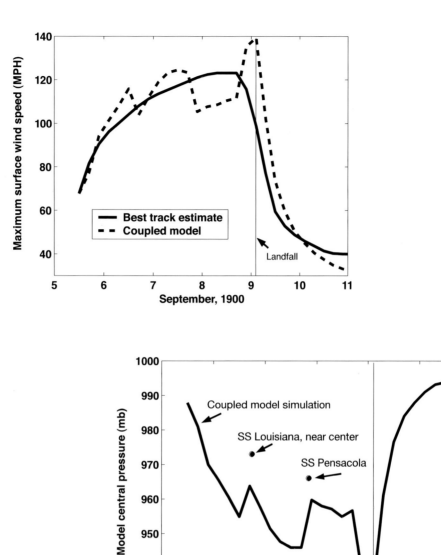

1876, killing more than a hundred thousand people, were impossible in places like Galveston, owing to the very gradual shoaling of the waters approaching the coast. Cline subscribed to this idea and had also theorized that the very low land inshore of Galveston Bay would absorb any flooding and prevent waters from rising very far. These beliefs led him to state in 1891 that "it would be impossible for any cyclone to create a storm wave which could materially injure the city," an opinion that probably helped defeat a proposal by a group of concerned residents and businessmen to build a seawall around Galveston. Yet twice in the previous sixteen years, the port of Indianola, just 150 mi southwest of Galveston, had been wrecked by hurricanes. The second storm, in 1886, obliterated the city with a large storm surge, and it was never rebuilt. Why Cline chose to ignore these events in favor of mere theory is anyone's guess.

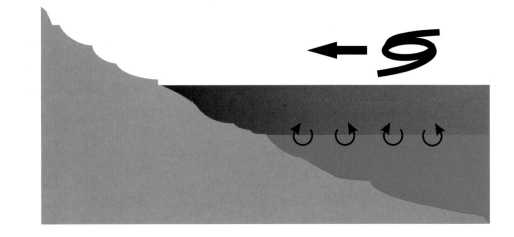

Figure 13.3: A hurricane offshore continually cools the warm layer of seawater near the surface, by stirring in cold water from below. But as the storm approaches land, the shoaling sea bottom cuts off the supply of cold water, and the surface waters remain warm, intensifying the hurricane.

▶

Thus it was that Cline viewed the rising waters and crashing surf with ambivalence on that Saturday morning. There are conflicting stories of what he did during the morning and early afternoon of September 8. According to his own recollections, he ignored regulations and issued hurricane warnings, persuading six thousand people to take shelter in the central city. On the other hand, there is ample evidence that the Galveston Weather Bureau office, under his command, never suggested that a dangerous storm was on the way, and few who survived the storm remember any warnings.

Later research, motivated in part by the Galveston tragedy, showed that storm surges are amplified, not diminished, by gradually shoaling waters. And as the storm approached the city from the southeast, an even deadlier effect of the shallow water offshore came into play.

Figure 13.2 shows the maximum wind speeds and minimum surface pressure in a computer simulation of the hurricane, from the time it left the coast of Cuba, and compares it with independent estimates made by the National Weather Service. Note that late on September 8 (Universal Time), the modeled storm began to intensify rapidly, leading to winds of about 60 m/s (140 mph) at the time of landfall. These wind speeds are consistent with estimates made at the time by several ships just offshore and by Weather Bureau observers, although the anemometer used to measure wind speeds blew away two hours before the storm reached its peak in Galveston, after registering winds of 100 mph. The official reconstruction is somewhat more conservative. The bottom panel of Figure 13.2 shows the corresponding record of central surface pressure simulated by the model and compares it to measurements made from two ships and an estimate made at Galveston. The dip in central pressure just before landfall corresponds with the increase in wind speed, and the lowest central pressure is in good agreement with the official estimate of 931 mb. In the days leading up to landfall, the modeled central pressure is somewhat lower than the observations by the SS *Louisiana* and the SS *Pensacola*, but there is no way of knowing just how close to the center these ships were.

The observations made on that fateful night in 1900 cannot tell us whether the great hurricane really did intensify rapidly just before landfall. But it is easy to discover just why the model storm did so, as illustrated in Figure 13.3. Like most tropical ocean waters, the Gulf of Mexico is surmounted by a relatively thin layer of warm water, extending down to between 20 m and perhaps 50m (50–150 ft). Beneath this warm layer, the water is much cooler. As discussed in Chapter 12, hurricanes stir up this cold water as they pass, cooling the surface and thereby reducing the storm's intensity, which is sensitive to the ocean surface temperature. But as a hurricane approaches shore, the sea floor gradually rises to a point where only the warm water layer remains. Beyond this point, with no cold water to stir up, the sea surface remains warm and the hurricane intensifies. The smaller the slope of the sea floor, the further offshore this point lies and the more time the storm has to intensify. It is this effect that accounts for the sudden intensification of the simulated Galveston storm just before landfall.

It was as if all the chance meteorological and oceanographic processes churning within the chaos of nature had conspired with human failing to bring about the destruction of Galveston. The storm took aim for a point just southwest of the city, insuring that it would be exposed to the strongest winds and highest storm surge. During the afternoon of Saturday, September 8, the wind continued to increase as the water rose and the barometer fell. Confused residents, unaware of the magnitude of the catastrophe that was about to befall them, sought refuge from the rising water and wind in their homes or in the sturdier buildings downtown. The storm surge itself arrived around 7:30 in the evening but was presaged by a more gradual rise of the ocean. In Cline's words:

> The water rose at a steady rate from 3 P.M. until about 7:30 P.M., when there was a sudden rise of about four feet in as many seconds. I was standing at my front door, which was partly open, watching the water, which was flowing with great rapidity from east to west. The water at this time was about eight inches deep in my residence, and the sudden rise of 4 feet brought it above my waist before I could change my position. The water had now reached a stage 10 feet above the ground at Rosenberg Avenue (Twenty-fifth street) and Q Street, where my residence stood. The ground was 5.2 feet elevation, which made the tide 15.2 feet. The tide rose the next hour, between 7:30 and 8:30 P.M., nearly five feet additional, making a total tide in that locality of about twenty feet. These observations were carefully taken and represent to within a few tenths of a foot the true conditions.

Houses on the waterfront quickly succumbed to the onslaught of high water, enormous waves, and winds that may have been as high as 60 m/s (140 mph). Huge timbers from disintegrating buildings became lethal flying missiles, damaging still-intact structures downwind and killing anyone in their path. Behind this aerial bombardment, an expanding juggernaut of debris began advancing inland, pushed by the storm surge and waves. Few structures survived this nightmare of wind, water, and debris. Cline himself, along with his brother, wife, and three children and about 45 others who had sought refuge with them, were trapped in their house as it collapsed around them. Only 18 of the 50 survived. Cline and his children were afloat for more than three hours, clinging to debris while trying to avoid the timbers flying around them at high velocity. But his wife, Cora, was among the thousands who perished that night.

The lowest measured pressure at Galveston, 964 mb (28.47"), was recorded at 8:30 P.M., as the storm center passed by a short distance to the southwest. Winds shifted rapidly from east to south, allowing the full force of the storm surge to enter Galveston Bay. Earlier that afternoon, all telegraph communication to the mainland had been cut off, so that no one outside the city had any real inkling of the disaster until well into the following day.

Figure 13.4: Aftermath of the Galveston Hurricane of 1900.

▲

The scene of devastation that presented itself to survivors when the sun rose on Sunday, September 9, beggars description. The wreckage of some 3,600 buildings stretched as far as the eye could see, while the air filled with the stench of decaying corpses. Yet early reports reaching Houston that five hundred people had died were considered gross exaggerations.

Gradually, during the following week, news of the catastrophe filtered to the outside world and relief efforts were organized. Among the first on the scene was the aging Clara Barton, founder of the American Red Cross, who had tended the injured on Civil War battlefields 35 years earlier. Relief workers, faced with an overwhelming number of corpses, had no choice but to collect them in piles and burn them. By night, the red glow of numerous funeral pyres, together with the awful stench and the returning heat and humidity, produced a facsimile of Hell, through which stunned survivors roamed in silence.

Although Galveston was eventually rebuilt, this time with a substantial seawall, it never recovered its place in Texas history and was quickly overshadowed by Houston, some miles inland but connected to the Gulf by a canal. The world quickly forgot the tragedy at Galveston, and, until interest in the event was revived by Erik Larson's *Isaac's Storm*, few Americans were taught about it. When asked, people often state that the 1906 San Francisco Earthquake was the nation's worst natural disaster.

Time will tell how much we have learned from the events leading up to that terrible September night.

WASN'T THAT A MIGHTY STORM

A folk song. Recorded by Sin-Killer Griffin and His Congregation in 1934.

Library of Congress, Washington, D.C.

Wasn't that a mighty storm
Wasn't that a mighty storm with water
Wasn't that a mighty storm
that blew the people all away

Now that town had her seawall
to keep the water down
but the high tide from the ocean
was water over the town

The prophet give them warning
you'd better leave this place
they never thought of leaving
as death looked them in the face

The trains they were loaded
with people leaving town
the track give way on the ocean
the trains they went on down

Then like a cruel master
as the wind began to blow
rode out on the trail of horses
said death let me go

Now death in nineteen hundred
that was fifteen years ago
you throw the storm at my mother
with you she had to go

Now death your hands is icy
you got them on my knees
you done carried away my mother
now come back after me

The trees fell on the island
the houses give away
some people strived and drownded
some died their separate way

The lightning blazed as kindlin'
the thunder began to grow
the wind it began blowin'
the rain began to fall

The sea it began growin'
the ship could not stand
I heard the captain cryin'
please save a drownding man

Wasn't that a mighty storm
Wasn't that a mighty storm with water
Wasn't that a mighty storm
that blew the people all away

After the Hurricane. Oil
painting by American artist
Marsden Hartley, 1938.

An awful Tempest mashed the air—
The clouds were gaunt, and few—
A Black—as of a Spectre's Cloak
Hid Heaven and Earth from view.

The creatures chuckled on the Roofs—
And whistled in the air—
And shook their fists—
And gnashed their teeth—
And swung their frenzied hair.

The morning lit—the Birds arose—
The Monster's faded eyes
Turned slowly to his native coast—
And peace—was Paradise!

—Emily Dickinson (1830–1886)

14   Genesis

He sprang from the womb of some wild cloud,
And was born to smite and slay:
To soar like a million hawks set free
And swoop on his ocean prey!

—William Hamilton Hayne, "A Cyclone at Sea"

The Tropics are primed to produce hurricanes. Like a coiled spring, the atmosphere-ocean system is poised to unleash huge reservoirs of energy. But fortunately for us, hurricanes are unusual: only about 85 tropical cyclones develop globally each year, of which roughly half go on to become full-fledged hurricanes. Even during the peak of hurricane season, a vacation in the Tropics is a good bet: the chances of being affected by a hurricane in any given year are miniscule.

Tropical cyclones develop in three principal belts, as shown in Figure 14.1. The first

Figure 14.1: Origin points of tropical cyclones over a 30-year period.

▶

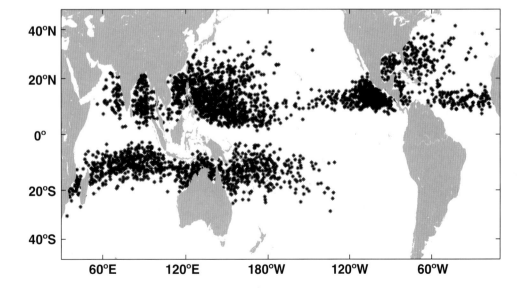

Figure 14.2: The average
number of tropical cyclones
per month in the Northern
and Southern Hemispheres.

▶

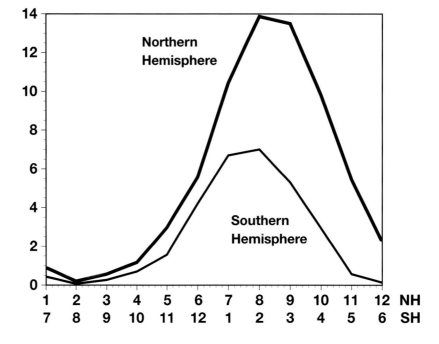

Figure 14.3: Evolution of the
maximum wind speed in two
computer hurricane simula-
tions. The first, given by the
solid curve, begins with a
vortex whose maximum wind
speed is 12 m/s (27 mph), and
the second uses an initial
vortex with maximum winds
of only 2 m/s (5 mph).

▶

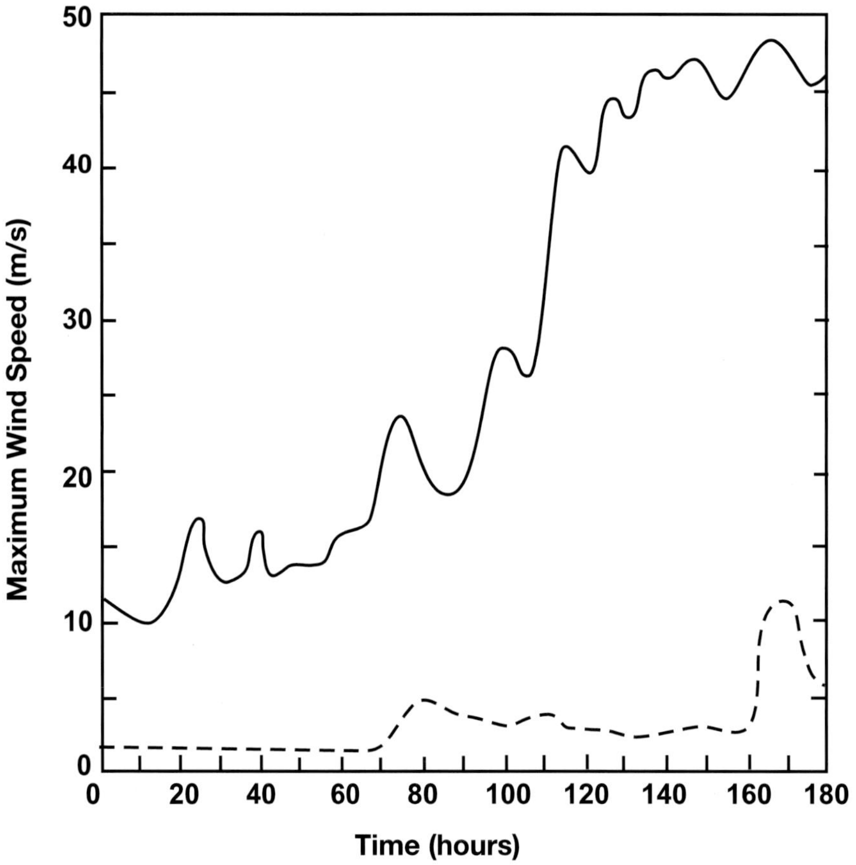

extends from the western tropical Atlantic across the Caribbean Sea and Gulf of Mexico, then continues westward through the tropical eastern North Pacific. The second begins in the central North Pacific and stretches westward across the South China Sea, the Bay of Bengal, and the Arabian Sea. The third belt extends from the central South Pacific across the Coral Sea and the south Indian Ocean, all the way across Madagascar to Africa.

Hurricanes tend to follow the sun, peaking in the summer and autumn and all but vanishing during the early spring (Figure 14.2). Typhoons can develop over the western North Pacific during any month of the year, though they are rare in winter.

The problem for research scientists is not why hurricanes develop, but why they hardly happen.

A clue is provided by recent computer simulations of hurricanes. Figure 14.3 shows the evolution of the peak wind speed in two simulations. The first begins with a moderate vortex whose maximum winds near the surface are about 12 m/s (27 mph). After several days of gestation, the model storm develops into a full-fledged hurricane.

The second is identical to the first, but starts with a very weak vortex, with maximum winds of only about 2 m/s (4–5 mph). This storm never develops, confirming what forecasters have long known: hurricanes do not develop spontaneously; they always require a trigger. This is quite unlike most other storms, such as winter storms, that develop spontaneously when the conditions are right.

An old-fashioned gas-powered lawn mower serves as an analogy. There it sits in your garage, with a tank full of gas and everything in working order, but except in low-budget sci-fi films, it does not start up by itself. Similarly, an automobile engine or a jet turbine must be started by using an electric motor to crank up the engine to some critical speed, whereupon, if all goes well, the engine itself starts working.

This presents at least three important scientific problems: Why do hurricanes have to be triggered? Under what circumstances can hurricanes form? And, in nature, what serves as the trigger?

The answer to the first question has to do with the humidity of the tropical atmosphere. In the normal state of the Tropics (see Chapter 4), warm air close to the surface rises in deep cumulonimbus clouds, losing most of its water content to precipitation. Thus when the air sinks back down toward the surface, it is quite dry. Mixing with shallower cumulus clouds, along with partial reevaporation of falling precipitation, moistens the air as it sinks, but it is still very dry. Figure 14.4 shows a typical profile of relative humidity in the tropical atmosphere. Above a thin layer near the surface, the humidity ranges between 60 and 70 percent. At some altitudes, it is even drier than this.

A decrease in humidity with altitude is also associated with a decrease in the energy content of the air, as can be seen in the bottom panel of Figure 2.6. This middle-level dryness presents a significant barrier to the development of hurricanes. To understand this, suppose cumulonimbus clouds clump together, for one reason or another, as shown in Figure 14.5. To support the updrafts inside the clouds, air near the sea surface must flow into the clouds (red arrows). But more updrafts give more rain, and the partial reevaporation of the rain drives stronger downdrafts (blue arrows), which bring low-energy air from the

Figure 14.4: A typical profile of relative humidity in the tropical atmosphere.

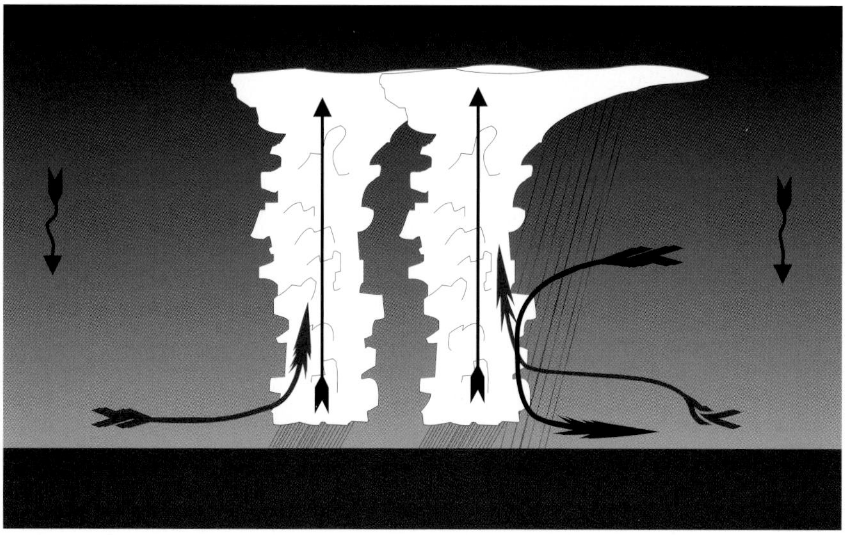

▶

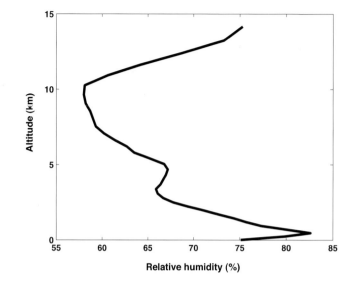

Figure 14.5: The congregation of tropical convective clouds leads to a local concentration of both updrafts (red arrows) and downdrafts (blue arrows). The downdrafts bring low energy down to the surface from middle levels of the atmosphere.

▶

Figure 14.6: The evolution of the maximum wind speed in three computer simulations of tropical disturbances. In the first (solid curve), the simulation is started with a vortex whose maximum wind speed is 18 m/s (40 mph). The second simulation (dashed curve) is like the first but starts from a maximum wind speed of only 2 m/s (5 mph). The third simulation (dash-dot curve) is like the second, but the atmosphere has first been humidified in a column 160 km (100 mi) across, centered on the storm.

▶

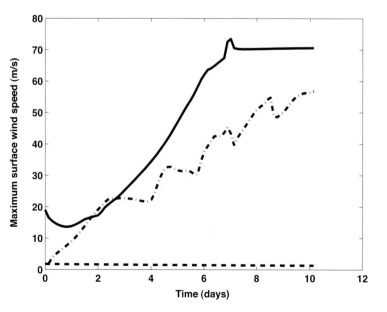

middle atmosphere down to the surface. This is just the opposite of what happens in a mature hurricane, where strong winds increase the energy content of the air near the surface (see Chapter 10). The nascent cloud cluster dies away.

If a hurricane is to develop, the "resistance" offered by the dry air at middle levels of the atmosphere must be overcome. This can happen when an atmospheric disturbance moistens the middle atmosphere over a region at least 80 km (50 mi) in diameter. Figure 14.6 compares three computer simulations using a simple hurricane model. The solid line shows the evolution of the maximum wind speed in a standard simulation beginning with a moderately strong vortex. The dashed line shows the same kind of simulation, but starting with a very weak vortex. (These two simulations are similar to those shown in Figure 14.3.) Finally, the dash-dot line shows a simulation starting from the same weak vortex, but where we have first humidified the whole atmosphere in a pillar about 100 mi across, centered on the storm. With no resistance to overcome, the weak vortex quickly amplifies.

The exact process by which the middle atmosphere becomes humidified remains enigmatic. There are no doubt several routes to genesis, and a variety of conditions can make the atmosphere more or less susceptible to genesis; these were first identified by William Gray some 30 years ago.

One factor is the humidity itself. The more humid the middle atmosphere, the less it will have to moisten for genesis to occur. In general, the air will be relatively humid where it is rising on a large scale, as in the ascending branches of the Hadley Circulation, while air descending on a large scale tends to be dry.

A major inhibitor of genesis is the presence of vertical wind shear, defined as the change in direction and/or speed of the wind with altitude. (See Chapter 12 for a precise definition.) If there is no wind shear, a nascent tropical system just moves along with the wind (which, in the absence of shear, has the same speed and direction at all altitudes). No air flows through the storm. But if wind shear is present, the storm tries to move along at an average wind speed, and at some altitudes wind must blow through the storm. This storm-relative flow can import dry air from outside the cloud cluster, destroying the humid column of air that is needed for genesis. For this reason, wind shear is inimical to genesis.

Hurricanes are rotary storms, and they must develop in an atmosphere that has some "spin." Because the earth rotates, even an atmosphere at rest has some spin. Imagine looking down on the North Pole from space: the earth and its atmosphere and oceans are rotating counterclockwise about an axis pointing straight up from the surface. Now imagine looking down at the earth from a position directly above the equator. Here there is no spin around an axis pointing straight up. Thus the component of spin increases from zero at the equator to a maximum value at the North Pole (and a minimum value at the South Pole).[9] Notice in Figure 14.1 that no hurricanes develop at the equator. (Yet they can get quite close. The record is Typhoon Vamei of December 27–28, 2001, which developed near Indonesia at a north latitude of merely 1.5°.)

In addition to the spin created by the rotating earth, the atmosphere can rotate with respect to the earth. We measure this rate of rotation by a quantity called *vorticity*.[10] When the rotation is in the same sense as the local vertical

---

[9] Technically, this rate of spin varies as the sine of the latitude.

[10] The curl of vector velocity. In most usage, we mean the local vertical component of this vector.

component of the earth's rotation, the vorticity is *cyclonic*; when it is in the opposite sense, it is *anticyclonic*. (Thus air having cyclonic vorticity is rotating counterclockwise with respect to the earth in the Northern Hemisphere, and clockwise in the Southern Hemisphere.) Gray recognized that, all other things being equal, genesis is more likely where air has cyclonic vorticity.

Finally, genesis cannot occur if there is no fuel in the tank; thus we require substantial potential intensity (see Chapter 10), as shown, for example, in Figure 10.3.

Gray developed an index that accounts for all these factors. Here we show an updated version of his index, applied to monthly mean climatological conditions of the atmosphere and ocean. Figure 14.7 shows, at each location, the maximum value of the index that occurs over the 12 months of the year. This can be compared to the observed locations of genesis shown in Figure 14.1. The major belts of tropical cyclogenesis are well captured by the index.

Although meteorologists are good at estimating whether the atmosphere is primed for hurricanes, we are less successful in identifying the triggers that determine exactly when and where storms will develop. What kinds of disturbances trigger hurricanes?

Over the North Atlantic and Caribbean, many hurricanes develop from atmospheric disturbances known as *African easterly waves.* These may be 1,500 km (1,000 mi) in dimension and are evident as oscillations in the general east-to-west flow of the atmosphere that prevails over the tropical North Atlantic in summer. They develop over sub-Saharan Africa and move off the west coast at a rate of about one every three days. Most continue westward across the tropical Atlantic and Caribbean, and may travel as far

west as the Gulf of Mexico or the eastern North Pacific. Many are associated with changes in wind speed and/or direction, and also with changes in rainfall. But few develop into tropical cyclones.

African easterly waves owe their existence to the Sahara Desert. In summer, the desert floor and the first 3–5 km (2–3 mi) of air above it get exceptionally hot. But the air over the Atlantic to the south remains relatively cool, so that between the Atlantic and the Sahara, the temperature actually increases northward. This is associated with easterly winds that increase with altitude. The resulting easterly jet, 3–5 km (2–3 mi) above the surface, is very unstable and breaks down into easterly waves.

Figure 14.8 shows a sketch of a typical easterly wave passing over the eastern North Atlantic after emerging from Africa. The map is meant to represent conditions about 3 km (2 mi) above the surface. There is high pressure to the north and east, and lower pressure to the south and west; the thin black curves show lines of equal pressure (isobars). The air tends to flow along isobars, with low pressure to the left. The easterly wave appears as a trough in the pressure field, and winds have a counterclockwise curvature near the trough (i.e., they have cyclonic vorticity). Easterly waves tend to be strongest about 3 km (2 mi) above the surface: they diminish both above and below that level. A gentle wind shift is sometimes noticed at the surface as the trough axis passes, but often there is little evidence of the wave in the surface winds. The trough itself moves westward and sometimes a little northward as well, traveling about 800 km (500 mi) in a day.

Besides influencing the winds, easterly waves have a distinct effect on the distribution of

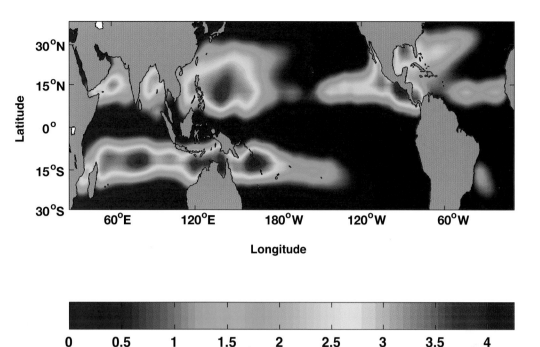

Figure 14.7: Map showing the maximum monthly value of a genesis index over the course of a year. This index has been calculated using monthly average values of wind shear, relative humidity, and potential intensity; the units of the index are arbitrary.

▶

clouds and rain. Depending on the meteorological circumstances, tropical rain showers may be heavier on the west or the east side of the trough axis, with unusually clear weather on the other side of the axis. By the time the trough reaches the central Atlantic, most of the bad weather usually lies east of the trough, while clear skies prevail to the west, as depicted in Figure 14.9.

As an easterly wave approaches, the normal tropical rain showers yield to mostly clear skies, with only a few "trade cumuli" dotting the skyscape. Then, as the wind shifts from northeasterly to southeasterly, heavy downpours commence and may last, on and off, for a day or so. Then conditions return to normal, with mostly sunny skies interrupted only by the odd rain shower.

How do these relatively benign easterly waves develop into hurricanes? And why do so few of them do so? These issues remain enigmatic and controversial. We observe that occasionally, especially in late summer and early fall,

the amount of convection associated with a particular wave increases, and winds near the surface evolve from the typical wavy pattern of an easterly wave into a closed circulation. A tropical depression is born. If conditions remain favorable, the depression may develop further into a tropical storm and, later, into a full-blown hurricane. The satellite photo in Figure 14.10 shows tropical systems in various stages of development over the North Atlantic. An easterly wave is just emerging from the west coast of Africa. (Note the dust blowing westward from the Sahara.) Over the central Atlantic, an earlier easterly wave is just beginning to develop into Tropical Storm Gustav. Further west, just north of the Windward Islands, Hurricane Fran is intensifying. And west of Bermuda, full-blown Hurricane Edouard is moving northward over the western Atlantic.

Easterly waves are only one of many ways of triggering tropical cyclones. A hurricane will sometimes form on an ordinary cold front that manages to penetrate the Tropics. Thunderstorm

Figure 14.8: Sketch of a typical African easterly wave, as it might appear on a map of pressure and airflow about 3 km (2 mi) above the surface. The black curves show isobars—lines of equal pressure—and the blue arrows show the airflow. The dashed black line marked "trough" shows the location of the wave axis, where there is a pressure trough and the airflow is cyclonic.

▶

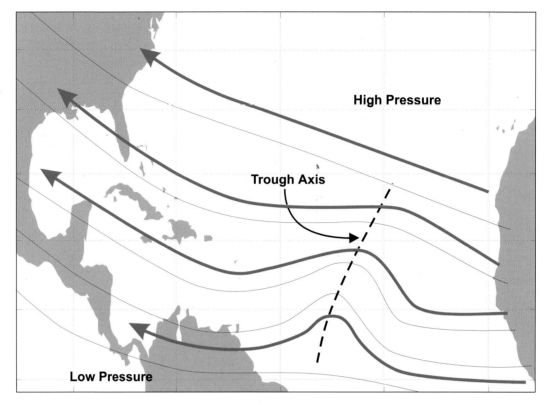

complexes that form over land occasionally drift out over tropical oceans and are transformed into tropical cyclones. The physical processes that create ordinary low-pressure systems over land, the kind that bring rain to middle- and high-latitude regions every few days, may sometimes trigger tropical cyclones if they occur over warm enough water. Over the western North Pacific, another type of easterly wave, quite different from the African variety, is often the culprit.

Although some aspects of the transformation of atmospheric disturbances into tropical cyclones are relatively well understood, the general problem of tropical cyclogenesis remains, in large measure, one of the great mysteries of the tropical atmosphere.

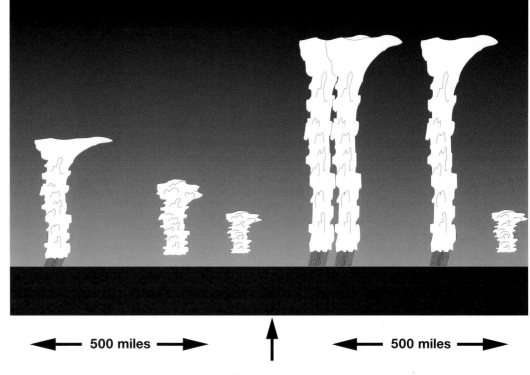

Figure 14.9: Sketch of the distribution of clouds and rain showers around an easterly wave, whose axis is in the center of the diagram. This distribution is characteristic of waves traversing the central and western North Atlantic.

▶

← 500 miles →   **← 500 miles →**

Trough Axis

Figure 14.10: Satellite photo of the tropical North Atlantic on August 31, 1996, showing various stages in the evolution of tropical disturbances. An easterly wave is just emerging over the ocean from western Africa. Further west, another such wave has developed into a tropical depression that will become tropical storm Gustav. The wave before that has become Hurricane Fran, northeast of the Virgin Islands, and the preceding wave is now fully developed Hurricane Edouard.

▶

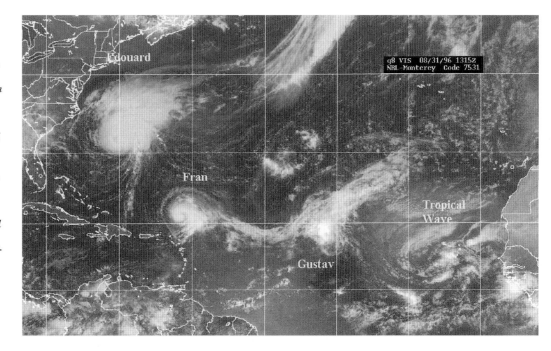

T he nakedness of a hurricane's truth is not revealed at once. You think you are seeing it within an hour after the wind comes, but your experience of the pitiless majesty of nakedness is enlarged from moment to moment. It must have been toward three in the morning that we saw the real thing. We had been bailing hard and had the boat clear of water to the battens. Tavi and I were lying side by side. He gripped my arm, and at the same moment I was aware of something more coming. It was that strange, immediate awareness of deeper menace one had during the war, when, as one lay under an intensive bombardment, scores and hundreds of the enemy's guns were suddenly added to those already in action. Intensity was intensified beyond all conceivable limits. So here it was.

The island, what little remained of it, disappeared from view. Raising my head to the level of the gunwale and shielding my eyes with my hand, I looked sideways along what had been the beach. There was nothing to be seen—nothing: no church, no trees, no sign of any sort to show that land had ever been there.... Then came a deluge of rain that made the others seem light by comparison.... Presently, for a few moments, the moon was shining in all but cloudless sky.

Moonlight...the full moon...what suggestions of peace, of serenity, are in the words! There can be nothing more beautiful in nature than a coral island, on a windless night, under the light of the full moon, but I leave you to imagine the desolation of the scene we now beheld. I looked first toward the church, and where it had stood there was nothing but the endless procession of combers. No vestige of it remained above the waste of moonlit water. The whole of the village islet was like one of those great mid-ocean shoals so feared by mariners, except that there was still evidence that land had been there. Hundreds of palms were down, but others yet stood, with men, women and children in them. I should never have believed that the coconut palm had such resilient strength. The stems of those that remained were bent in what seemed impossible arcs, but the sea was their great enemy, washing away their holding ground so that the wind could take them. So many had gone that I could now see for the first time one of the old purau trees that stood near the church. It was a superb old tree, with a trunk four feet in diameter, and seemed a contemporary of the island itself. I could make out several people clinging to what remained of it, but at that distance it was impossible to recognize them. The other purau tree that had stood near by was gone.

—Charles Nordhoff and James Norman Hall, from *The Hurricane*

15　　Miami, 1926

Late that night, in absolute darkness, it hit, with the far shrieking scream,
the queer rumbling of a vast and irresistible freight train.
　　—Marjory Stoneman Douglas

Man has never learned why the Almighty sees fit to destroy cities.
　　—Editorial, *Miami Tribune*, September 19, 1926

In the summer of 1926, Miami, Florida, was beginning to suffer the economic and social hangover that often follows periods of rapid economic expansion and land speculation. According to legend, the city was founded after Julia Tuttle, a divorcée from Chicago and early settler, sent a bouquet of orange blossoms to the great railway tycoon Henry Flagler, to show him that south Florida had not suffered the devastating frost that had just affected the northern part of the state. Flagler had resisted earlier entreaties to extend his Florida East Coast Railway south from West Palm Beach, but the prospect of year-round good weather finally decided him. He extended his railroad, and the city of Miami was incorporated on July 18, 1896.

The lure of warm weather, beautiful beaches, and cheap land drew many settlers to south Florida, which, by 1926, boasted a population of nearly three hundred thousand. In the three years between 1920 and 1923, the population of Miami itself doubled. During 1925 alone, more than $60 million worth of buildings were erected in the city. But by early 1926, the boom began to fizzle. Credit became scarce as stories of scams and corruption began to circulate, scaring away potential settlers and investors alike.

Into this teetering society moved the most intense hurricane to affect Florida since it was first settled.

The sixth tropical cyclone of 1926 began as a classic Cape Verde storm, passing north of Puerto Rico on September 15. Thereafter it largely escaped detection until it passed over the Bahamas during Friday the seventeenth. As it approached Miami, the textbook signs of an oncoming storm were largely absent: Weather Bureau observers noticed no telltale cloud formations, nor did they observe particularly large ocean swells, thanks to the barrier provided by the Bahama Islands. The

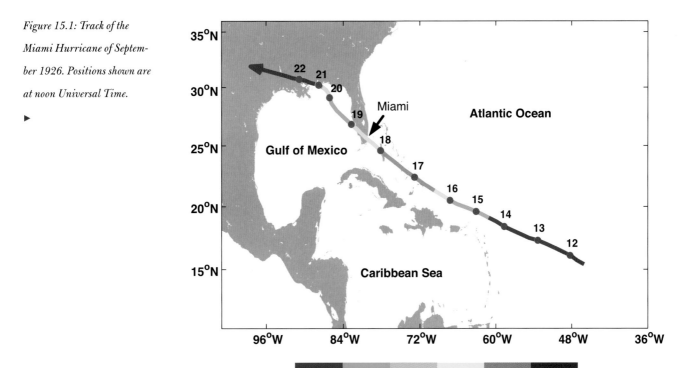

morning edition of the *Miami Herald* ran a four-inch column describing a storm in the Bahamas but assuring readers that it would miss Florida. A few Miamians went to the beach expecting to see impressive surf, but they were disappointed. Yet by Friday afternoon, the *Miami News* reported that storm warnings were in effect, and destructive winds were expected that evening. Few residents took notice. The official U.S. Weather Bureau surface map, drawn at 8 P.M. on Friday evening, makes it clear that the Bureau's Washington, D.C., office had no clue about the strong hurricane bearing down on Florida.

Hurricane flags were not hoisted until 11:25 on the evening of the seventeenth, long after most residents had gone to bed. The winds, which had been quite gentle until late in the evening, increased rapidly after midnight, as the barometer plummeted, and by dawn were gusting to well over 45 m/s (100 mph). Many residents awoke in the middle of the night to the sounds of howling winds, tearing awnings, breaking glass, and rising water. Almost no one had experienced a hurricane, and most did not know what to do. Their ignorance of hurricanes became an especially lethal liability early that Saturday, when the eye of the storm passed right over downtown Miami around 6 A.M.

Records from the Weather Bureau office, directly in the path of the storm, indicate that hurricane-force northeast winds diminished rapidly and became light and variable for 35 minutes, beginning at 6:10 A.M. Dazed citizens emerged from hiding, believing the storm was over. The Bureau chief, Richard Gray, shouted at people near his office to go back to shelter, that the storm was not over; but they paid him no heed. At 6:45 A.M., the eastern eyewall struck with even greater force, killing many who had taken to the streets and washing into Biscayne Bay many cars that had started across the causeway connecting Miami Beach to downtown Miami. Few understood that hurricanes have calm

Figure 15.2: Miami at the
height of the 1926 hurricane.

▲

eyes. At 7:30 A.M., an anemometer in Miami Beach recorded a wind velocity of 57 m/s (128 mph), while a storm surge of 2.5–4.5 m (8–15 ft) inundated Miami Beach and downtown Miami, sending a bore up the Miami River. Buildings that had been weakened by the initial onslaught now collapsed.

In the words of Richard Gray, "The intensity of the storm and the wreckage that it left cannot be adequately described. The continuous roar of the wind; the crash of falling buildings, flying debris, and plate glass; the shriek of fire apparatus and ambulances that rendered assistance until the streets became impassable; the terrifically driven rain that came in sheets as dense as fog; the electric flashes of live wires have left the memory of a fearful night in the minds of the many thousands that were in the storm area."

Yet the hurricane was not finished. As it passed south of Lake Okeechobee in south central Florida, 45 m/s (100 mph) northeast winds blew a wall of lake water into the town of Moore Haven, drowning as many as three hundred people.

Miami lay in ruins. Several hundred people were dead, and the storm caused about $150 million in damage (roughly equivalent to $1.7 billion today). In a single day, the storm had wrecked more property than had been built in two years, plunging the city into depression three years before the rest of the country followed suit.

In contemplating the magnitude of the destruction caused by the 1926 hurricane, it is sobering to know that it was only the twelfth most intense hurricane to strike the United States from the time record keeping began until the time of this writing. Hurricane-force winds affected Miami for 24 hours, a record of duration unmatched today. It has been estimated that were an identical storm to take the same path through Miami today, it would cause $87 billion in damages, compromising the entire U.S. insurance industry. While warnings have improved immensely since 1926, the example set 66 years later by Hurricane Andrew shows that our penchant for building inadequate structures near hurricane-prone coasts has greatly increased our economic vulnerability to natural disaster.

▲

Hurricane Force. Oil
painting by French artist
Owanto (Yvette Berger),
1990.

A THROAT of thunder, a tameless heart,
    And a passion malign and free,
He is no sheik of the desert sand.
    But an Arab of the sea!

He sprang from the womb of some wild cloud,
    And was born to smite and slay:
To soar like a million hawks set free,
    And swoop on his ocean prey!

He has scourged the Sea till her mighty breast
    Responds to his heart's fierce beat,
And has torn brave souls from their bodies frail
    To fling them at Allah's feet.

Possessed by a demon's lust of life,
    He revels o'er wrecks and graves,
And hurtles onward in curbless speed,—
    Dark Bedouin of the waves.

—William Hamilton Hayne
    (date unknown), "A Cyclone at Sea"

16   Death and Transfiguration

The morning lit—the Birds arose—
The Monster's faded eyes
Turned slowly to his native coast—
And peace—was Paradise!

        —Emily Dickinson

When the core of a hurricane makes landfall, its death usually comes quickly. Suddenly bereft of their source of power—evaporation of warm ocean water—the now unopposed frictional forces rapidly diminish the speed of the winds, especially near the surface. Wind speeds diminish exponentially, losing half their landfall value in about 7 hours, 75 percent in 15 hours, and 90 percent in a day. Thus even a 65 m/s (150 mph) hurricane will be reduced below hurricane strength in about 7 hours, and after a day all that remains is a light breeze. Over mountainous terrain, death can come even faster.

When hurricanes pass over swamps instead of dry ground, however, they can suck enough heat out of the swampy water to slow their decay. Even a foot of standing water holds enough heat to noticeably slow a storm's decay, as shown in Figure 16.1. A swamp 1 m (3 ft) deep can sustain a marginal hurricane indefinitely.

Hurricane Andrew (Chapter 31) provides a good example of how swamps can fuel storms. After slamming into southern Florida with 74 m/s (165 mph) winds, Andrew crossed directly over the Everglades, a vast wetland south of Lake Okeechobee. Though over Florida for less than a day, it picked up enough heat from the Everglades to slow its dissipation, possibly leading to a more intense storm in Louisiana several days later (Figure 16.2).

Rainfall also generally diminishes after landfall, though not as quickly as the winds, because storms retain a great deal of water vapor and cloud as they move inland.

Although most hurricanes quickly dissipate after making landfall or moving over cold ocean water, it is not uncommon for storms to partially regenerate as they enter middle and high latitudes, even if they remain over land or cold water. Meteorologists call this *rejuvenation*, and it may be accompanied by increasing winds and rains, or merely by a decrease in the rate of dissipation.

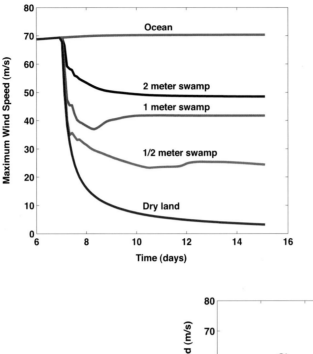

Figure 16.1: Decay of storm winds after landfall, assuming that the maximum wind speed at landfall is 68 m/s (150 mph). Each curve shows the decrease of wind speed with time in storms making landfall at seven days, on the time scale shown on the bottom axis. The bottom curve is for storms moving inland over flat, dry ground, while the other curves represent storms moving over swamps of various depths. The top curve, provided for reference, is for a storm that remains at sea.

▶

Figure 16.2: Observed evolution of the maximum wind speed in Hurricane Andrew (blue curve) compared to two computer simulations. The first assumes southern Florida to be dry land, while the second accounts for the Everglades.

▶

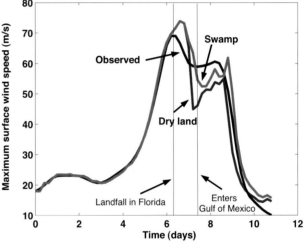

A notorious case of rejuvenation is that of Hurricane Hazel of 1954. Hazel struck North Carolina with 62 m/s (140 mph) winds and then accelerated to more than 18 m/s (40 mph), reaching Ontario less than a day after landfall (Figure 16.3). The winds, rather than dying down rapidly after landfall, leveled off at around 45 m/s (100 mph), doing extensive damage through North Carolina, Virginia, and Pennsylvania, and as far north as Toronto.

The New England Hurricane of 1938 is another example of a rejuvenating storm that caused severe wind damage well inland (Chapter 21).

Rejuvenation occurs when hurricanes or their remnants interact with middle- and high-latitude weather systems—the familiar high- and low-pressure systems shown on television weather reports. These systems, known as *extratropical cyclones* and *anticyclones,* are responsible for most day-to-day variations in weather outside the Tropics. They derive their energy from horizontal temperature contrasts, and so are strongest in winter, when such contrasts are largest.

For extratropical cyclones and anticyclones to develop, ascending air must be warmer than descending air. Such an arrangement lowers the center of gravity of the atmosphere, since cold

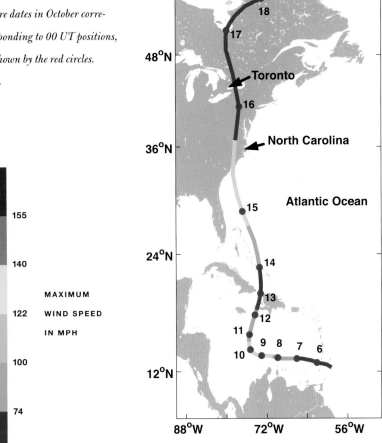

Figure 16.3: Track of Hurricane Hazel. Numbers are dates in October corresponding to 00 UT positions, shown by the red circles.

▶

air is denser than warm air, converting potential energy to kinetic energy, just as in a roller coaster accelerating down an incline.

In a volume of air that is not rotating, such a conversion will occur spontaneously: any horizontal temperature gradient will result in air motion that reduces and finally eliminates the temperature gradient. (By the same token, any tilt of water in a glass will lead to sloshing of the water until the surface is level.) But the rotation of the atmosphere greatly complicates the energy conversion, and under some circumstances can prevent it altogether, allowing horizontal temperature gradients to exist in a stable equilibrium.

Extratropical cyclones and anticyclones develop by breaking down this balance, allowing colder air to sink and warmer air to rise. The actual physics of this process were not well understood until the late 1940s. Even today, many research papers are devoted to this subject. The cyclones and anticylones are not efficient enough to eliminate horizontal temperature gradients; otherwise, the temperature at the poles would be the same as in the Tropics. In reality, the average pole-to-equator temperature difference is roughly half of what it would be if the atmosphere did not move at all.

During rejuvenation, a hurricane or its remnant encounters the stronger horizontal temperature contrasts and associated wind systems prevalent in middle and high latitudes. When this happens, it can trigger the sinking of cold air and rising of warm air that converts potential energy into kinetic energy, thereby increasing the storm's own intensity. In other words, it can begin to tap the reservoir of energy associated with horizontal temperature contrasts, and it can do this long after it has lost its oceanic energy source. This transformation of a tropical cyclone into an extratropical

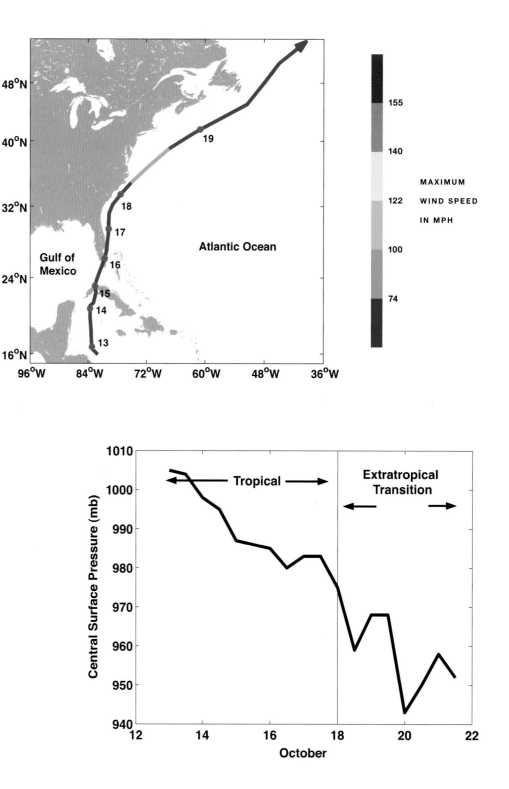

Figure 16.4: Track of Hurricane Irene. Numbers are dates in October corresponding to 00 UT positions, shown by the red circles.

▶

Figure 16.5: Evolution of the central pressure of Irene, showing both tropical and extratropical phases.

▶

cyclone is called *extratropical transition*. Under some circumstances, storms can tap both energy sources simultaneously.

Rapid changes in storm intensity and structure often accompany extratropical transi-

tion. Typically, the circularly symmetric cloud system of the tropical cyclone gradually transforms to the highly asymmetric cloud geometry that characterizes extratropical cyclones, with most of the clouds and precipitation located

Figure 16.6: Satellite image of
Hurricane Irene undergoing
extratropical transition.

▶

poleward of the surface low-pressure system, and along curving fronts that usually emanate eastward and southwestward from the storm center (northwestward in the Southern Hemisphere). Rainfall can be extremely heavy, and winds can exceed hurricane strength in transitioning systems.

Hurricane Irene of 1999 is a good example of extratropical transition. As shown in Figure 16.4, Irene began in the Caribbean and moved northward across Cuba and southern Florida.

Then, as it rapidly accelerated northeastward into the North Atlantic, Irene underwent a spectacular transition into an extratropical cyclone. In fact, the strongest winds developed after the transition, when the storm was extratropical. Figure 16.5 shows a history of Irene's central surface pressure, a measure of the storm's intensity (with lower pressure associated with stronger winds).

A satellite image taken about the time Irene began its transformation is shown in Figure 16.6. The eye has filled with cloud and, in addition to the spiral of cloud around the core, a large, thick mass of cloud is developing northwest of the center where the storm's counterclockwise flow is lifting warm air up over cold air in New England and eastern Canada. Another cloud band extends eastward from north of the storm center, along a developing warm front where the storm's circulation drives warm, moist air up and over cold air to the north. A day later, this transformation resulted in a cloud pattern typical of an extratropical cyclone, with little evidence that it had once been a hurricane.

Although some ex-hurricanes are reborn as extratropical cyclones, others are not, even when they move poleward into regions of large horizontal temperature contrasts. One discriminating factor is the chance encounter of the tropical cyclone with extratropical weather systems. Rejuvenation is more likely when an upper-level extratropical cyclone approaches the tropical system from the west. (For example, a powerful upper-level cyclone overtook the New England Hurricane of 1938 from the west.) Rejuvenation also seems to be more likely over the ocean, especially when the storm is moving rapidly from warm to cold water. When this happens, the air near the surface quickly cools. The cool, dense air forms a kind of insulating cushion that protects the vortex aloft from the ravages of surface friction, as can be seen by comparing two computer simulations of hurricanes (Figure 16.7). In the first (top panel), the hurricane makes landfall at 180 hours, after which the winds diminish rapidly. In the second (bottom panel), the hurricane, at 180 hours, suddenly moves over ocean that is

$10°C$ colder than before. Note that while the winds near the surface drop off even more rapidly than in the landfalling storm, the winds aloft decay much more slowly. The strength of the vortex aloft may give it an edge when and if it subsequently interacts with middle-latitude weather systems.

Finally, the enormous, thin anticyclones at the tops of hurricanes (Chapter 2) can also interact with middle-latitude weather systems in complex ways, even when extratropical transition is weak or absent. As a tropical cyclone moves into middle latitudes, these anticyclones are typically swept downstream (usually to the east) by powerful winds in the upper atmosphere, where they can trigger the development of extratropical fronts and cyclones. An especially notorious case was the gale that devastated southern England in October 1987. Although the storm was incorrectly called a "hurricane" by British media (it did have hurricane-force winds), research suggests that it was greatly enhanced by an anticyclone that swept eastward from the top of a marginal hurricane that had occurred many days earlier near Florida.

While some aspects of the interaction between tropical and extratropical cyclones are beginning to be understood, much of the relationship remains enigmatic, posing another challenge to science.

Figure 16.7: These two panels show the decline of hurricane wind speeds at each altitude in computer simulations of a hurricane moving over dry land (top) and over cold ocean water (bottom), starting at 180 hours. The shading shows the wind speed (in mph) according to the scales at right. The simulations use the numerical model developed by Rotunno and Emanuel (1987).

▶

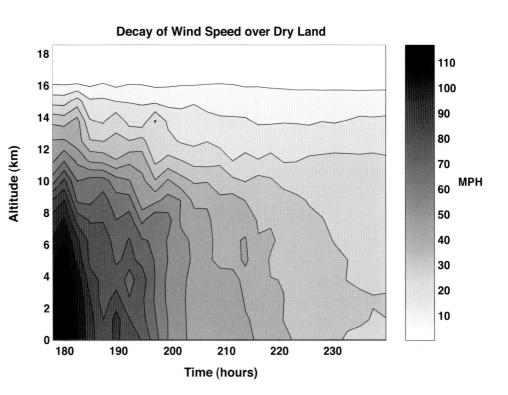

The Young America in a Storm. Oil painting by American artist James Edward Buttersworth, ca. 1855.

Happy the man who, safe on shore,
    Now trims, at home, his evening fire;
Unmov'd, he hears the tempests roar,
    That on the tufted groves expire:
Alas! on us they doubly fall,
Our feeble barque must bear them all.

Now to their haunts the birds retreat,
    The squirrel seeks his hollow tree,
Wolves in their shaded caverns meet,
    All, all are blest but wretched we—
Foredoomed a stranger to repose,
No rest the unsettled ocean knows.

While o'er the dark abyss we roam,
    Perhaps, with last departing gleam,
We saw the sun descend in gloom,
    No more to see his morning beam;
But buried low, by far too deep,
On coral beds, unpitied, sleep!

But what a strange, uncoasted strand
    Is that, where fate permits no day—

No charts have we to mark that land,
    No compass to direct that way—
What Pilot shall explore that realm,
What new Columbus take the helm!

While death and darkness both surround,
    And tempests rage with lawless power,
Of friendship's voice I hear no sound,
    No comfort in this dreadful hour—
What friendship can in tempests be,
    What comfort on this raging sea?

The barque, accustomed to obey,
    No more the trembling pilots guide:
Alone she gropes her trackless way,
    While mountains burst on either side—
Thus, skill and science both must fall;
And ruin is the lot of all.

    —Phillip Freneau (1752–1832),
    "The Hurricane"

17 Their Eyes Were Watching God: *San Felipe* and the Okeechobee Disaster of 1928

One word describes it. It was Hell! A raging
inferno of rolling, swirling waters, of shrieking,
demoniac winds, of lashing rain and of darkness,
black and absolute. There were no atheists that
night on the shores of Okeechobee!

—Lawrence E. Will

Lake Okeechobee (a Seminole word meaning "Big Water") covers almost seven hundred square miles of south central Florida, the second-largest body of fresh water that is entirely contained within U.S. borders. Too big to see one end from the other, it is nevertheless remarkably shallow: one can wade through most of it. It is fed from the north by several freshwater streams, most notably the Kissimmee River, and until the early years of the twentieth century, water drained out its south end through the Everglades, an enormous swamp covering most of Florida south of Okeechobee.

To recover potentially valuable farmland, the government began an ambitious project to drain the Everglades in 1910. Workers built canals to drain water eastward and southward from Lake Okeechobee, lowering the water level and drying out parts of the Everglades. An earthen dike 5 to 8 ft high was built around the lake's south end to control the floods that often followed heavy rains. To cultivate the new land, farmers brought in thousands of migrant workers, many from the Bahamas. Several new towns developed on the shores of the lake. Belle Glade, on the southeastern shore, was incorporated in April 1928, and while it was home to thousands of sharecroppers and migrant workers, it had only 209 registered voters.

During most of the Everglades reclamation, south Florida was spared a major storm. But the Miami Hurricane of 1926 revealed the great susceptibility of the new lakeside towns to flooding from Okeechobee. Passing just south of the lake, the hurricane's 45 m/s (100 mph) east-northeast winds piled Okeechobee's water up against its southwestern banks, breaching the dike and flooding the town of Moore Haven, drowning hundreds of people (Chapter 15).

Two years later, a far worse disaster was visited on the south shores of Okeechobee.

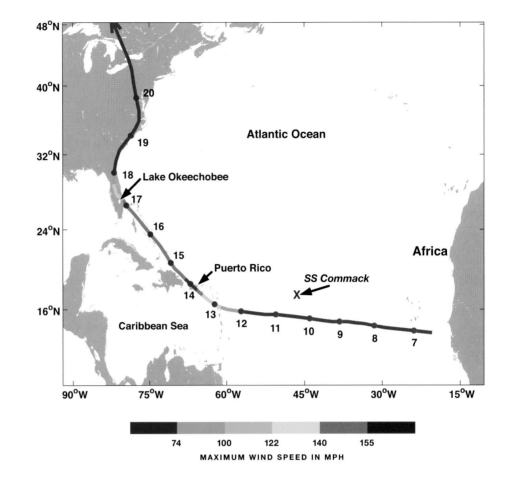

The storm originated in an African easterly wave and was first reported in the eastern Atlantic by the crew of the SS *Commack* at latitude 17° N, longitude 48° W. It first made landfall at Pointe á Pitre, Guadeloupe, around noon on September 12, recording a surface pressure of 940 mb (27.76"). The storm destroyed almost every building on the island and killed more than five hundred people. On the nearby islands of St. Kitts and Montserrat, another hundred perished. At about 11 A.M. on the thirteenth, the storm passed directly over the SS *Matura*, southwest of St. Croix in the Virgin Islands, registering a central pressure of 931 mb (27.50"). Later that day, it passed into Puerto Rico, making landfall near the coastal village of Arroyo, 32 mi southeast of San Juan. Anemometers in San Juan recorded sustained winds of 65 m/s (150 mph) before blowing away, and as much as 75 cm (30") of rain fell in the central mountains. The devastation in Puerto Rico was appalling. More than 19,000 buildings were razed. Tin roofs, popular at that time, became horrible airborne instruments of decapitation in the high winds. Torrential rains resulted in severe flooding. When it was over, 1,500 Puerto Ricans were dead and 284,000 had lost their homes. The storm, which became known as the "San Felipe Hurricane" after the saint's day on which it struck the island, remains Puerto Rico's worst hurricane disaster.

Puerto Rico is mountainous, and detailed records of more recent storms show that winds weaken rapidly as hurricanes pass over the island. It is likely that the San Felipe storm was considerably diminished as it emerged from the northwest coast of the island late on September 13, though

Figure 17.2: Evolution of the maximum wind speed in the San Felipe Hurricane of 1928. The blue curve shows the official estimate, and the red curve shows the results of a computer simulation.

▶

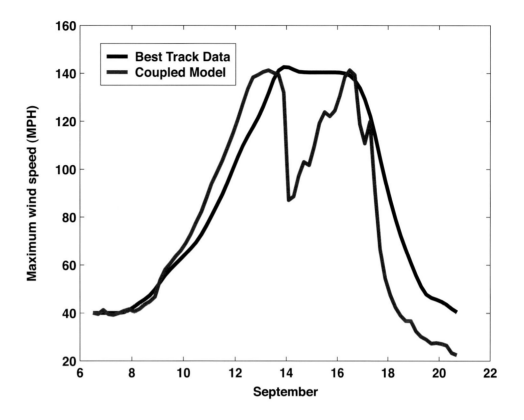

historical data from the National Hurricane Center show no such diminution (Figure 17.2). To estimate the effect of Puerto Rico's terrain on the San Felipe Hurricane, I ran a computer model of hurricane intensity using the best estimated track of the storm. The result, also shown in Figure 17.2, shows the expected influence of the mountains, with maximum wind speeds diminished to about 40 m/s (90 mph) on the fourteenth. This is consistent with observations of 45 m/s (100 mph) winds and 951 mb (28.09") central pressure by the time the storm reached Nassau, in the Bahamas, on the fifteenth.

As the hurricane approached the east coast of Florida, it gathered strength after the blow dealt by Puerto Rico, aided by the deep warm waters of the Gulf Stream. But the U.S. Weather Bureau mislocated the storm and failed to issue timely warnings, believing it would turn north and stay out at sea. By the time it hit an unprepared Palm Beach on the evening of September 16, its central pressure had dropped to an astonishing 929 mb (27.44"), the lowest central pressure recorded up to that time in the United States. San Felipe's sustained winds of 63 m/s (140 mph) drove a 3.3 m (11 ft) storm surge over the island of Palm Beach, killing at least 25; washing out the major coastal highway A1A; and destroying countless homes, including some of the most expensive mansions in the country.

Residents of the towns on the south shore of Lake Okeechobee had heard radio reports of a severe hurricane in the Caribbean and the Bahamas, but most were assured by the forecasts that it would miss Florida. The lake was already high, owing to unusually heavy rains over the previous two months and to appeals by fishermen and residents along the drainage canals to keep water levels up. Now, driven by an increasing northerly wind, the water levels on the south shore began to rise even

Figure 17.3: Destruction in the aftermath of the San Felipe Hurricane.

▶

higher. By the time the eye passed directly over the lake late on the evening of September 16, the earthen dikes had given way, and a wall of water swept inland, driven by 55 m/s (120 mph) winds. The primitive houses, many of which had been erected in the last two or three years, were instantly crushed, and the debris formed a rolling wall whose advance destroyed everything in its path. Those who escaped this onslaught sought refuge in trees, where many were killed by poisonous water moccasins also fleeing the rising water. Several people were swept miles into the Everglades; a few managed to find their way back to civilization, but many never made it. Virtually everyone who had elected to ride out the storm on the lake islands perished, as did a group of migrant workers who, ignoring warnings that strong winds would return, ventured out during the lull accompanying the passage of the eye. Some 274 women and children survived aboard two barges that miraculously rode out the storm. Others managed to clamber to the roof of a Belle Glade hotel, the only building in that town to come through relatively intact.

News of the disaster at Okeechobee was slow to reach the outside world. The nearest substantial city was West Palm Beach, some 45 mi to the east; it was itself reeling from the storm, and all telegraph lines were down. A full day and a half later, on Tuesday the eighteenth, headlines around the United States screamed news of a disaster on Florida's east coast, unaware of the far greater tragedy inland. When refugees began to trickle into West Palm Beach, the extent of the disaster slowly dawned.

To this day, the number of people who lost their lives on the shores of Okeechobee is known only very roughly. Migrant workers were not included in censuses, and at the time, state politics were dominated by northern Floridians, many of whom took a dim view of immigrants in the south. In the words of Florida's attorney general, Fred Davis, "It is mighty hard to get people in other parts of the State interested in whether they [the workers of south central Florida] perish or not." This callousness toward the welfare of the migrant workers, most of whom were black, contributed to the lack of an accurate death toll. The Red Cross initially estimated that 2,300 had died in the storm, but owing to intense pressure from officials who feared that such a large number would scare away tourists and investors, this figure was revised downward to the absurdly precise figure of 1,836, a number still repeated in contemporary literature. A simple count of the number of people reported to have been buried after the storm exceeds 2,400.[11] There can be little doubt that this was the second-worst hurricane disaster is U.S. history, exceeded only by the Galveston storm of 1900, and the second- or third-worst natural disaster, comparable to the Johnstown Flood of 1889, which killed more than 2,200. For many years afterwards, farmers cultivating land south of the lake would come across human skeletons, the last vestiges of that terrible storm.

Nine years later, in 1937, the horrors of the Okeechobee hurricane as experienced by migrant workers and their families were recounted by Zora Neale Hurston in her novel *Their Eyes Were Watching God*:

> So she was home by herself one afternoon when she saw a band of Seminoles passing by. The men walking in front and the laden, stolid women following them like burros. She had seen Indians several times in the 'Glades, in twos and threes, but this was a large party. They were headed towards the Palm Beach road and kept moving steadily. About an hour later another party appeared and went the same way. Then another just before sundown. This time she asked where they were all going and at last one of the men answered her.
>
> "Going to high ground. Saw-grass bloom. Hurricane coming."
>
> Everybody was talking about it that night. But nobody was worried. The fire dance kept up till nearly dawn. The next day, more Indians moved east, unhurried but steady. Still a blue sky and fair weather. Beans running fine and prices good, so the Indians could be, must be, wrong. You couldn't have a hurricane when you're making seven and eight dollars a day picking beans. Indians are dumb anyhow, always were. Another night of Stew Beef making dynamic subtleties with his drum and living, sculptural, grotesques in the dance. Next day, no Indians passed at all. It was hot and sultry and Janie left the field and went home.
>
> Morning came without motion. The winds, to the tiniest, lisping baby breath had left the earth. Even before the sun gave light, dead day was creeping from bush to bush watching man.
>
> Some rabbits scurried through the quarters going east. Some possums slunk by and their route was definite. One or two at a time, then more. By the time the people left the fields the procession was constant. Snakes, rattlesnakes began to cross the quarters. The men killed a few, but they could not be missed from the crawling horde. People stayed indoors until daylight. Several times during the night Janie heard the snort of big animals like deer. Once the muted voice of a panther. Going east and east. That night the palm and banana trees began that long distance talk with rain. Several people took fright and picked up and went in to Palm Beach anyway. A thousand buzzards held a flying meet and then went above the clouds and stayed.
>
> One of the Bahaman boys stopped by Tea Cake's house in a car and hollered. Tea Cake came out throwin' laughter over his shoulder into the house.
>
> "Hello Tea Cake."
>
> "Hello Lias. You leavin', Ah see."
>
> "Yeah man. You and Janie wanta go? Ah wouldn't give nobody else uh chawnce at uh seat till Ah found out if you all had anyway tuh go."

[11] Reported in the excellent book *Killer 'Cane* by Robert Mykle.

"Thank yuh ever so much, Lias. But we 'bout decided tuh stay."

"De crow gahn up, man."

"Dat ain't nothin'. You ain't seen de boss man go up, is yuh? Well all right now. Man, de money's too good on the muck. It's liable tuh fair off by tuhmorrer. Ah wouldn't leave if Ah wuz you."

"Mah uncle come for me. He say hurricane warning out in Palm Beach. Not so bad dere, but man, dis muck is too low and dat big lake is liable tuh bust."

"Ah naw, man. Some boys in dere now talkin' 'bout it. Some of 'em been in de 'Glades fuh years. 'Tain't nothin' but uh lil blow. You'll lose de whole day tuhmorrer tryin' tuh git back out heah."

"De Indians gahn east, man. It's dangerous."

"Dey don't always know. Indians don't know much uh nothin', tuh tell de truth. Else dey'd own dis country still. De white folks ain't gone nowhere. Dey oughta know if it's danger-ous. You better stay heah, man. Big jumpin' dance tuhnight right heah, when it fair off."

Lias hesitated and started to climb out, but his uncle wouldn't let him. "Dis time tuhmor-rer you gointuh wish you follow crow," he snorted and drove off. Lias waved back to them gaily.

"If Ah never see you no mo' on earth, Ah'll meet you in Africa...."

After a while somebody looked out and said, "It ain't gitting no fairer out dere. B'lieve Ah'll git on over tuh mah shack." Motor Boat and Tea Cake were still playing so everybody left them at it.

Sometime that night the winds came back. Everything in the world had a strong rattle, sharp and short like Stew Beef vibrating the drum head near the edge with his fingers. By morn-ing Gabriel was playing the deep tones in the center of the drum. So when Janie looked out of her door she saw the drifting mists gathered in the west—that cloud field of the sky—to arm them-selves with thunders and march forth against the world. Louder and higher and lower and wider the sound and motion spread, mounting, sinking, darking.

It woke up old Okeechobee and the monster began to roll in his bed. Began to roll and complain like a peevish world on a grumble. The folks in the quarters and the people in the big houses further around the shore heard the big lake and wondered. The people felt uncomfortable but safe because there were the seawalls to chain the senseless monster in his bed. The folks let the people do the thinking. If the castles thought themselves secure, the cabins needn't worry. Their decision was already made as always. Chink up your cracks, shiver in your wet beds and wait on the mercy of the Lord. The bossman might have the thing stopped before morning anyway. It is so easy to be hopeful in the day time when you can see the things you wish on. But it was night, it stayed night. Night was striding across nothingness with the whole round world in his hands.

A big burst of thunder and lightning that trampled over the roof of the house. So Tea Cake and Motor stopped playing. Motor looked up in his angel-looking way and said, "Big Massa draw him chair upstairs."

"Ah'm glad y'all stop dat crap-shootin' even if it wasn't for money," Janie said. "Ole Massa is doin' His work now. Us oughta keep quiet."

They huddled closer and stared at the door. They just didn't use another part of their bodies, and they didn't look at anything but the door. The time was past for asking the white folks what to look for through that door. Six eyes were questioning *God*.

Through the screaming wind they heard things crashing and things hurtling and dashing with unbelievable velocity. A baby rabbit, terror ridden, squirmed through a hole in the floor and squatted off there in the shadows against the wall, seeming to know that nobody wanted its flesh at such a time. And the lake got madder and madder with only its dikes between them and him.

In a little wind-lull, Tea Cake touched Janie and said, "Ah reckon you wish now you had of stayed in yo' big house 'way from such as dis, don't yuh?"

"Naw."

"Naw?"

"Yeah, naw. People don't die till dey time come nohow, don't keer where you at. Ah'm wid mah husband in uh storm, dat's all."

"Thanky, Ma'am. But s'posing you wuz tuh die, now. You wouldn't git mad at me for draggin' yuh heah?"

"Naw. We been tuhgether round two years. If you kin see de light at daybreak, you don't keer if you die at dusk. It's so many people never seen de light at all. Ah wuz fumblin' round and God opened de door."

He dropped to the floor and put his head in her lap. "Well then, Janie, you meant whut you didn't say, 'cause Ah never knowed you wuz so satisfied wid me lak dat. Ah kinda thought—"

The wind came back with triple fury, and put out the light for the last time. They sat in company with the others in other shanties, their eyes straining against crude walls and their souls asking if He meant to measure their puny might against His. They seemed to be staring at the dark, but their eyes were watching God....

They stepped out in water almost to their buttocks and managed to turn east. Tea Cake had to throw his box away, and Janie saw how it hurt him. Dodging flying missiles, floating dangers, avoiding stepping in holes and warmed on the wind now at their backs until they gained comparatively dry land. They had to fight to keep from being pushed the wrong way and to hold together. They saw other people like themselves struggling along. A house down, here and there, frightened cattle. But above all the drive of the wind and the water. And the lake. Under its multiplied roar could be heard a mighty sound of grinding rock and timber and a wail. They looked back. Saw people trying to run in raging waters and screaming when they found they couldn't. A huge barrier of the makings of the dike to which the cabins had been added was rolling and tumbling forward. Ten feet higher and as far as they could see the muttering wall advanced before the braced-up waters like a road crusher on a cosmic scale. The monstropolous beast had left his bed. The two hundred miles an hour wind had loosed his chains. He seized hold of his dikes and ran forward until he met the quarters; uprooted them like grass and rushed on after his supposed-to-be conquerors, rolling the dikes, rolling the houses, rolling the people in the houses along with other timbers. The sea was walking the earth with a heavy heel.

"De lake is comin'!" Tea Cake gasped.

Loss of HMS Blenheim *and* Java *in a Hurricane off Rodriguez.* Oil painting by British Artist Thomas Buttersworth, 1807.

One night came on a hurricane,
The sea was mountains rolling,
When Barney Buntline turned his quid,
And said to Billy Bowline:
"A strong nor'wester's blowing, Bill.
Hark! Don't you hear it roar now?
Lord help them! How I pities all
Unlucky folks on shore now.

"Foolhardy chaps that live in towns;
What dangers they are all in,
And now lie shaking in their beds
For fear the roof should fall in.
Poor creatures, how they envy us
And wishes, I've a notion,
For our good luck in such a storm
To be upon the ocean.

"And often, Bill, I have been told
How folks are killed, and undone,
By overturns of carriages,
By fogs and fires in London.
We know what risks all landsmen run,
From noblemen to tailors,
Then, Bill, let us thank Providence
That you and me are sailors"
          —William Pitt (d. 1840),
          "Sailor's Consolation"

18   Taking Aim: How Hurricanes Move

Out of the south cometh the whirlwind.

—Job 37:9

In the early days of hurricane prediction, forecasters felt lucky if they could locate a hurricane with reasonable certainty. Often, they first learned of storms after they had made landfall and already taken many lives. But thanks to telegraphy, reconnaissance aircraft, radar, and satellites, storms can now be located accurately. The problem is to forecast where the storms will go. Many of the great hurricane tragedies since the middle of the twentieth century can be traced to a failure to do this.

The physics of hurricane motion presents an interesting scientific problem. To a first approximation, hurricanes move with the airflow in which they are embedded, averaged through the depth of the storm. By "airflow in which they are embedded," we mean the large-scale flow of the tropical atmosphere that would exist in the absence of the storm. Look, for example, at Figure 8.5, which shows airflow near the surface of the Pacific Ocean. In the far western North Pacific, the swirl of a typhoon can be seen within the general east-to-west airflow.

Tropical meteorologists like to describe airflow using streamlines, curves on a weather map that are everywhere parallel to the wind. Under certain conditions, often approximately valid in the tropical atmosphere, we can build up complex flow patterns by adding the streamlines of a few simple flow patterns. Consider first a constant east-to-west flow, illustrated in Figure 18.1a. The streamlines are oriented east-west, and the spacing between them is constant, reflecting a constant wind speed.

Next, in Figure 18.1b, is a circular vortex, with counterclockwise winds. The spacing between the streamlines decreases inward toward the vortex center: the closer packed the streamlines, the stronger the wind. In Figure 18.1c, we have simply added the two streamline fields of (a) and (b) to get the total flow: a vortex embedded in a constant easterly wind. The resulting

Figure 18.1: Some idealized tropical flow patterns. A simple east-to-west flow is shown in (a), while a circular vortex, like a hurricane, is shown in (b). The two flows added together create the airflow pattern shown in (c), when the vortex is strong, and (d), when the vortex is weak.

▶

(a)

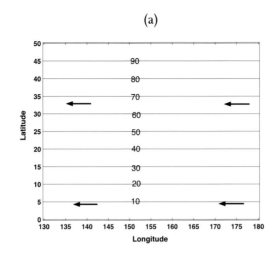

(b)

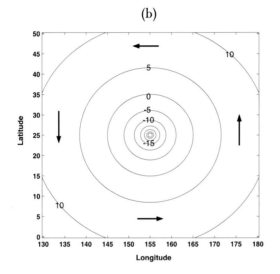

(c)

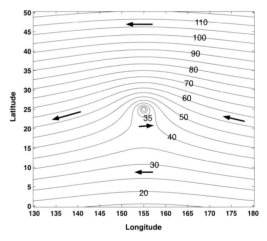

(d)

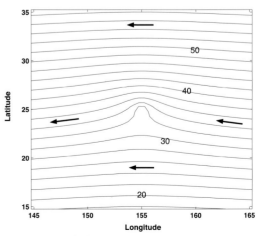

flow closely resembles the appearance of a tropical cyclone on a weather map. A few things to notice: 1) the streamlines are closer together, and therefore the wind is stronger, on the north side of the storm, where the vortex winds add to the background easterlies; 2) there is a "stagnation point" where the wind vanishes, near the middle of the map, and another one some distance to the south of the storm center; and 3) the true vortex center (b) is at 25°N, 155°E, but the circulation center in (c) is slightly south of this point. It is dangerous to use streamlines to locate a storm center, because the presence of a background wind shifts the position of the stagnation point relative to the true vortex center. This is particularly problematic when the vortex is weak, as shown in (d); here the vortex is too weak to cancel out the background easterlies, and there are no stagnation points. Meteorologists like to call such systems "open waves," but there is little or no physical significance to this term: we still have a vortex embedded in a flow.

The systems depicted in Figures 18.1c and 18.1d would move straight westward at the speed of the background flow. A trained forecaster looking at these maps would deduce that the flow consists of a constant wind with a vortex superimposed, and would be able to estimate the speed and direction of the background wind; this would make a good forecast of the storm's motion. A computer can easily extract the background wind from the flow in Figures 18.1c and 18.1d by averaging the wind around circles centered at the storm center, a procedure that eliminates the vortex flow but preserves the background wind.

The diagrams in Figure 18.1 show the flow at a single altitude, but hurricanes are usually embedded in background flows that change direction and speed with altitude. What should we use as the "steering level," i.e. the altitude at which the background wind provides a good estimate of storm motion? Experience shows that the best altitude is around 4–5 km (13,000–16,000 ft), roughly a third of the way through the depth of the storm. It is even better to average the background wind over the depth of the storm. One has to be careful, though, how this average is defined: we must account for the fact that air density decreases with altitude, so the flow at upper levels should not have as much weight as the low-level flow.[12] Fortunately, the flows in which most hurricanes are embedded usually do not change rapidly with altitude.

We now have a fairly simple picture: hurricanes just move with the background flow in which they live. We can use ordinary weather forecast models to predict the evolution of the background flow, and if we know where the storm is now, we can predict its motion for as long as we can predict the evolving background flow. We do not even have to simulate the internal structure of the hurricane: all we need do is add to our computer model a hypothetical entity that moves with the average background flow.

Even were this simple picture correct, there would be some very tough forecast scenarios. Consider the flow shown in Figure 18.2. It is broadly easterly at low latitudes and westerly at high latitudes, but there are two high-pressure systems located around 15°N. Midway between these centers is a special kind of stagnation point

[12] The density effect can be taken into account by averaging the flow at equal increments in pressure rather than equal increments of altitude. So, for example, averaging the flow at the surface, 1 km, 2 km, 3 km, etc. would give an incorrect density-weighted average, while averaging the flow at the 1,000 mb, 900 mb, 800 mb, etc. pressure levels would give the correct result.

Figure 18.2: Atmospheric flow
between two high-pressure
systems. The dot in the center
shows a "saddle point," and a
hurricane is located at the S.

▶

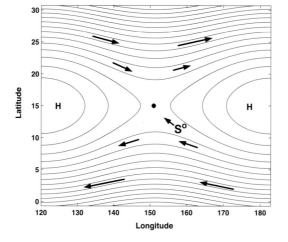

Figure 18.3: This diagram
of the Northern Hemisphere
shows how columns of air
displaced northward and
southward acquire spin
relative to the earth's surface.
The columns here extend from
the surface to the tropopause,
shown by the dashed curve.
The right side shows a column
of air displaced southward
and acquiring a counterclockwise spin, while on the left, a
column displaced northward
acquires clockwise spin.

▶

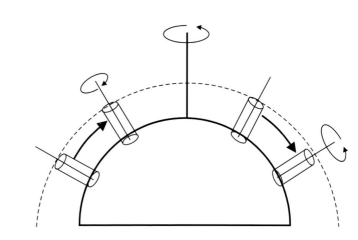

called a *saddle point*. Such points can wreak havoc with hurricane track forecasts. If a storm is embedded deep within either the low-latitude easterlies or the high-latitude westerlies, there is little problem predicting its path. But if a storm is approaching a saddle point, its trajectory becomes highly unpredictable. A storm approaching a saddle point is like a steel ball approaching a peg in a pinball machine: a slight perturbation in its approach can cause enormous differences in its subsequent trajectory. Of course, in the real world, the background flow evolves in time, so the saddle points are moving targets.

The general idea of predicting a hurricane's motion by using the background flow would work were it not for one crucial problem: hurricanes *change* their background flow. Up to now, we tacitly assumed that the vortex flow and the background flow evolve independently, but in the real world, each affects the other. This makes life more complicated but also more interesting.

How do hurricanes change the flow in which they are embedded? The best way to think about this is to recognize that the atmosphere tends to conserve a quantity called *absolute vorticity*, which is a measure of its spin about a vertical axis. Samples of air spin for two reasons: first, the air is rotating along with the earth, so even air at rest (with respect to the earth's surface) is spinning, as seen by an observer in space. Second, the air may also be rotating with respect to the earth.

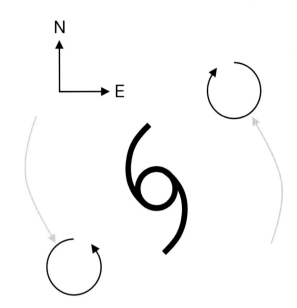

Figure 18.4: Air moving northward on the east side of a hurricane acquires clockwise spin; air moving southward west of the storm acquires counterclockwise spin.

▶

The total rate of spin, the absolute vorticity, is the sum of the spin of the earth plus the spin of the air with respect to the earth.

Air at rest at the equator has no absolute vorticity, because its axis is perpendicular to the earth's rotation axis. Air at rest at the North Pole has relatively large absolute vorticity, because its axis is parallel to the earth's axis. (The same is true of air at the South Pole, but its spin is in the opposite direction: its absolute vorticity is negative.)

In summary, the absolute vorticity of resting air is maximum at the North Pole, zero at the equator, and minimum at the South Pole.[13]

Absolute vorticity tends to be conserved in columns of air that extend from the surface to the tropopause (the top of the layer of air that experiences weather). So if a column of air initially at a high latitude is displaced southward, it will carry along its large absolute vorticity and therefore have greater vorticity than the air around it. This means it will now be spinning counterclockwise relative to the surface. Like-

wise, columns of air displaced northward acquire clockwise spin (Figure 18.3).

Air east of a hurricane is moving northward (in the Northern Hemisphere), while west of the storm it is moving southward. The northward-moving air acquires clockwise spin, and the southward-moving air acquires counterclockwise spin. Some of the clockwise spin east of the storm is swept around its north side, and some of the counterclockwise spin west of the storm is swept around to its south (Figure 18.4).

By this means, the hurricane creates two subsidiary vortices: a clockwise vortex to the northeast and a counterclockwise vortex to the southwest. Meteorologists call these *beta gyres*. They represent a third component to the flow, adding to the background flow and the hurricane itself. The total flow is the sum of all three components.

The flow around the subsidiary vortices "pushes" the hurricane toward the northwest (southwest in the Southern Hemisphere). This is called "beta drift," and its speed can be as

[13] The absolute vorticity of resting air actually varies as the sine of the latitude.

Figure 18.5: Rising air is replaced by converging airflow at low levels and produces diverging outflow aloft. The vorticity of the converging air increases, giving counterclockwise flow, while it decreases in the diverging air, giving clockwise flow. The sense of rotation is opposite in the Southern Hemisphere.

▶

Figure 18.6: Wind shear pushes the anticyclone at storm top off to one side. The low-level cyclone and the upper-level anticyclone then push each other in one direction, in this case, toward the north.

▶

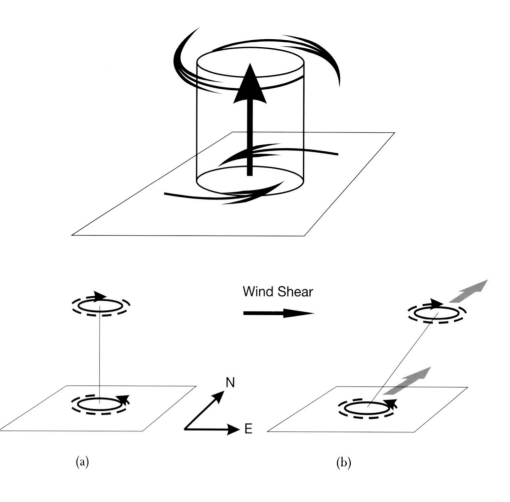

Wind Shear

(a)   (b)

much as 2 m/s (5 mph); failure to account for it can cause track forecasts to be off by as much as 120 mi in 24 hours. This is enough to determine whether a hurricane hits or misses a major city.

Beta drift causes hurricanes to move even in an atmosphere otherwise at rest. But there are other ways hurricanes can alter their own destiny. To understand these other ways, it is again helpful to use the concept of vorticity.

Although the vertically averaged vorticity tends to remain constant in columns of air, the vorticity can vary considerably from one altitude to the next within each column. A good example is the hurricane itself. Obviously, air is spinning very rapidly through most of the column containing the storm. But at the top of the storm, the air is rotating the opposite way (anticyclonically),

and although this air is not spinning very rapidly, it covers an area that is very large compared to the circulation near the surface. Averaged over the whole volume covered by the storm, the anticyclonic flow nearly balances the cyclonic flow lower down.

This variation of vorticity within columns is directly related to vertical air motion. Ascending air must be replaced by converging airflow below the level of maximum ascent, while above that level, the airflow must diverge, as illustrated in Figure 18.5. Conservation of angular momentum demands that the absolute vorticity must increase as air draws near the rotation axis, while it must decrease in air flowing away from the axis. (A classic demonstration of this effect is an ice skater retracting and extending her arms to spin

Figure 18.7: Pacific Typhoon Keith of 1997, intensifying toward supertyphoon status. Notice the particularly well developed cloud band west of Keith's core. The counterclockwise spin associated with this cloud band caused Keith to deviate to the north.

▶

up and spin down.) Thus in columns of ascending air, the absolute vorticity increases at lower levels and decreases at high levels. Divergence can never change the sign of the absolute vorticity. But remember that air with zero absolute vorticity (i.e. air that is not spinning as seen from an observer fixed in space) is moving with respect to the earth underneath it, which *is* spinning. So an observer on the earth sees such a column spinning anticyclonically (clockwise in the Northern Hemisphere).

Hurricanes act like chimneys spewing air with very small absolute vorticity into the upper atmosphere. If there is no background wind, this air tends to form an anticyclonic lens whose center is directly above the center of the low-level cyclone. But if there is some vertical wind shear,[14]

this lens is blown to one side of the surface cyclone. The flow around the anticyclone can then "push" the surface cyclone off course, as shown by Figure 18.6, while the surface cyclone pushes the anticyclone in the same direction. Recent research has shown that this effect can be as strong as beta drift.

Several other phenomena can alter hurricane movement. Unusually large cloud clusters sometimes develop on the storm's periphery, and these may develop cyclonic flow at low and middle levels of the atmosphere. This flow pushes the hurricane off course, as happened, for example, in Typhoon Keith (Figure 18.7). The unusually pronounced band of convection to the west of Keith's center developed its own cyclonic spin, which pushed Keith northward; likewise,

[14] See Chapter 12.

Figure 18.8: Typhoons Ione and Kristen of 1974, locked in a cyclonic dance.

▶

Keith's circulation pushed the enhanced convection to the south.

Sometimes, a cloud cluster itself grows into a tropical cyclone, or two cyclones develop simultaneously in close proximity, or approach each other after developing independently. When this happens, the mutual interaction of the two cyclones causes them to rotate around each other, with the weaker partner moving faster than the stronger one. This meteorological dance is known as the *Fujiwhara effect*, after the Japanese scientist who first described it. Figure 18.8 shows Pacific typhoons Ione and Kristen performing a cyclonic ballet in 1974.

Landfall can influence hurricane movement in several ways. Flat land does not much affect a storm's progress, though its rapid weakening will lessen self-steering effects such as beta drift. But high mountain ranges can cause complex distortions of the track. Sometimes hurricanes "jump" over mountains, reforming downstream at a location that may be far away from a simple extrapolation of its former track.

With all these effects in mind, let's have a look at the climatology of tropical cyclone tracks and a few case studies of individual storms.

Figure 18.9 shows tropical cyclone tracks over a ten-year period. Storms that originate deep within the Tropics almost always begin by moving westward and, to some extent, poleward, particularly in the Southern Hemisphere. Eventually, they either move over land or cold water

Figure 18.9: Tracks of tropical cyclones over ten years (1992–2001, Northern Hemisphere; 1991–2000, Southern Hemisphere).

▶

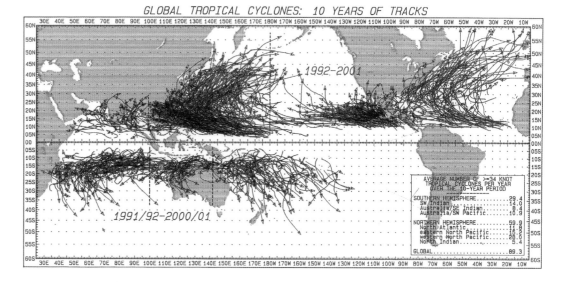

and die, or they recurve poleward and eastward. Most Atlantic storms and roughly half off all western North Pacific storms recurve, while eastern North Pacific tropical cyclones almost always move over cold water before they have a chance to recurve. Atlantic storms sometimes survive into high latitudes, because the high-latitude North Atlantic is warmer than other oceans at the same latitude.

Tropical cyclones occasionally follow bizarre tracks, performing loops, hairpin turns, and sharp curves. This can be a forecaster's nightmare, particularly if the storm is close to a populated coastline. A case in point is Hurricane Elena of 1985, whose track is shown in Figure 18.10. The storm developed from a cloud cluster that can be traced back to the Cape Verde Islands, and raced rapidly across the Atlantic. It traveled along the whole length of Cuba's north coast and entered the Gulf of Mexico late on the evening of August 28, when it was named. Sporting winds near 45 m/s (100 mph), it proceeded inexorably toward the Gulf coasts of Louisiana, Mississippi, Alabama, and the Florida panhandle, forcing authorities to evacuate large numbers of people. On the thirtieth it slowed dramatically,

and late in the day, while 200 mi from the Mississippi delta, Elena made a sharp right turn toward the west coast of Florida. As though confronted by a loose cannon on a heaving deck, surprised Floridians scrambled to safety as Elena crawled toward them. Meanwhile, the all-clear was sounded along the northern Gulf coast, prompting not only the return of residents, but an influx of Labor Day vacationers. Elena stalled only 50 mi from Florida, leaving residents and forecasters in suspense. Finally, the storm recommitted itself to Plan A and accelerated toward its original destination, prompting a second evacuation from Louisiana to the Florida panhandle. On Labor day, September 2, Elena made landfall near Biloxi, Mississippi. Economic losses from the storm exceeded $1 billion, and 1.5 million people had been evacuated, many of them twice.

Elena's pirouette was caused by an upper-atmospheric low-pressure trough progressing eastward across the U.S. mainland. South of the trough, middle-latitude westerlies extended southward to disrupt Elena's northward progress, turning her eastward. Then, as the trough passed by to the east, Elena was left to drift slowly in light winds. Finally, the prevailing southeast

Figure 18.10: Track of

Hurricane Elena of 1985.

▶

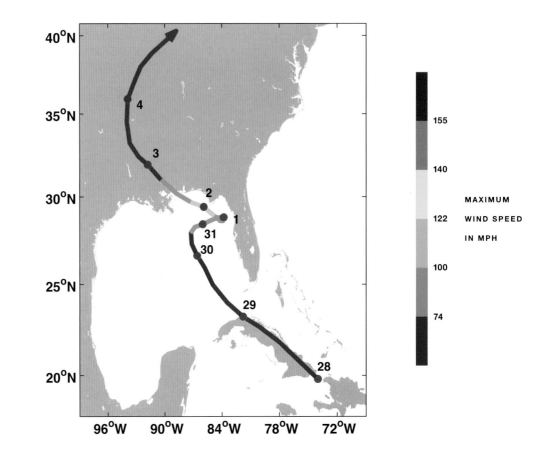

flow returned to guide Elena to the northern Gulf coast.

Although environmental influences on tropical cyclone movement can be complicated, today's computer models are in principle capable of accounting for them. As discussed in Chapter 29, computer forecasts of hurricane movement have improved steadily since they were first introduced around 1960. We can expect such improvements to continue, but they will eventually reach a limit dictated by the amount and quality of observations of the atmospheric environment in which hurricanes live, and by the inherently chaotic nature of the atmosphere (Chapter 29).

▲

The corvette *La Favorite* under command of French explorer Cyrille-Pierre-Theodore Laplace encounters a hurricane in the Indian Ocean, 1830. Engraving from Laplace's *Voyage autour du monde,* Paris, 1833–39.

I am the Rider of the wind,
 The Stirrer of the storm;
The hurricane I left behind
 Is yet with lightning warm;
To speed to thee, o'er shore and sea
 I swept upon the blast:
The fleet I met sail'd well, and yet
 'Twill sink ere night be past.

 —Lord Byron (1788–1824),
 "Fifth Spirit," from *Manfred,* Act I, Scene i

19   The Labor Day Hurricane of 1935

*Every day we hear discussion and dissertation upon the subject of the big wind, that
all-destroying tropical hurricane which has been fast approaching for the past month.
Where is the consarned thing? Aren't we important enough for the consideration
of even a decrepit, spavined has-been of a tropical zephyr?*

    —John Ambrose, correspondent for the *Key Veteran News*, August 31, 1935

*Indian Key absolutely swept clean, not a blade of grass, and over the high center of it
were scattered live conchs that came in with the sea, craw fish, and dead morays.
The whole bottom of the sea blew over it.*

    —Ernest Hemingway

T he most intense hurricane ever to affect the United States roared ashore in the middle Florida
Keys on the evening of Labor Day, September 2, 1935, wiping several islands clean of all vegetation and
buildings, and killing at least 423 people. Its central pressure of 892 mb (26.35") was the lowest
recorded in any storm to strike the United States and the lowest ever recorded in the Atlantic region
until 1988, when Hurricane Gilbert squeaked by with a central pressure of 888 mb. Among the casualties were many World War I veterans who had been sent to the Keys to help build a highway and, some
say, to prevent them from publicizing their grievances and embarrassing the administration of Franklin
D. Roosevelt. It so damaged the massive Key West Extension of Henry Flagler's Florida East Coast
Railway that it was never rebuilt. The storm killed more than half those residing in the Middle Keys.

    In 1935 the population of the Florida Keys was burgeoning, thanks to their deserved reputa-
tion as a tropical paradise and to their new accessibility, owing to the railway line built at the staggering cost of $49 million by the great tycoon Henry Flagler. Begun in 1905 and finished in 1912, the
system of bridges and viaducts linking the many islands was an engineering marvel. The finest Belgian
and German cement was used for the bridge piers, the steel came from Carnegie, and the marl road
bed fill was quarried and dredged from the Keys themselves. But at least two hundred workers lost
their lives in the project, many during the hurricanes of 1906, 1909, and 1910.

    The lure of the Keys, further enhanced by author Ernest Hemingway's presence in Key West,
drew an increasing number of settlers and tourists. In May 1935, the population of Key West was
12,470; and that of the Upper Keys was 673; but only 192 people inhabited the Middle Keys. Added
to these were the 750 veterans working on a new highway, designed to eliminate the many ferry runs
between the islands. Three years earlier, 20,000 vets, calling themselves the "Bonus Expeditionary

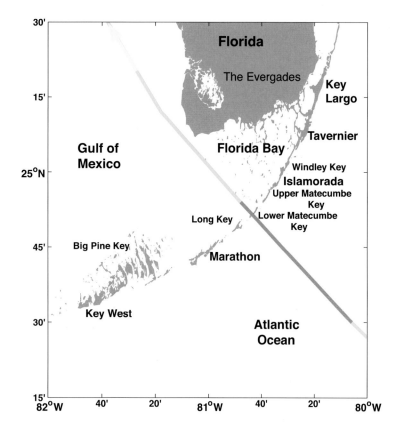

Figure 19.1: Map of the Florida Keys showing the track of the Labor Day Hurricane of 1935. Red indicates Category 5; orange Category 4; and yellow Category 3.

▶

Forces," had marched on Washington to complain that they had not been paid the bonus they had been promised. Partly to employ them and partly to get them out of the limelight, President Roosevelt put some of them to work on a new highway linking the Keys, beginning in 1934. They were housed in flimsy, temporary wooden buildings on Lower Matecumbe and Windley Keys (Figure 19.1. Windley Key is just northeast of Upper Matecumbe.)

The veterans were in rough shape. Shell-shocked from the war and largely neglected by the people for whom they had fought, many had taken to hard drinking and brawling. They were given free room and board and paid about $30 per month, but progress on the highway was slow. After one year, only two hundred feet of the roadway had been completed.

From the beginning there were worries about the veterans' safety. Among those concerned was Grady Norton, head of the U.S. Hurricane Warning Service in Jacksonville. At his urging, the Florida State Veterans Administration developed an evacuation plan that called for a train to take the vets to a temporary camp north of Miami in the event of a hurricane.

The Labor Day storm began somewhere east of the Bahama Islands on or about August 28 and proceeded slowly westward. At 1 P.M. on Saturday, August 31, a U.S. Weather Bureau advisory cautioned that a "tropical disturbance of small diameter but considerable intensity is central about 60 mi east of Long Island Bahamas apparently moving west-northwestward attended by strong shifting winds and probably gales near center. Caution advised southeastern Bahamas and ships in that vicinity." Late on the thirty-first, the fledgling storm passed onto the Great Bahama Bank, where the ocean is so shallow that the warm ocean mixed layer extends right to the sea floor. With no cold water to

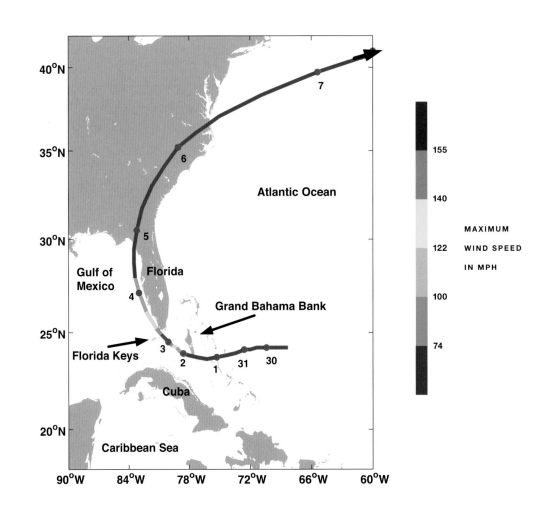

Figure 19.2: Track of the Labor Day Hurricane of 1935. Numbers show dates in August and September, corresponding to positions at 00 UT.

churn to the surface, the storm intensified rapidly, its slow forward movement giving it ample time to strengthen. At 9:30 on the evening of August 31, storm warnings were hoisted from Ft. Pierce to Miami, and the Weather Bureau warned residents of the Keys to take precautions. That evening, a severe windstorm affected Andros Island, the largest of the Bahamas, but no reports reached the mainland United States.

The Weather Bureau advisory issued at 9:30 A.M. Eastern Time, September 1, noted that the storm had attained hurricane status and projected it to continue moving west through the Straits of Florida and into the Gulf of Mexico by late that night or early Monday. Storm warnings were extended and now covered the whole coastline south from Ft. Pierce, around the southern tip of the state, and up to Ft. Myers on the west coast.

While the forecast had the storm moving too far south and too fast, the prediction was by no means unreasonable, given the available observations and the meteorological knowledge of the day. Moreover, the Weather Bureau was not taking chances: it posted warnings for all of south Florida. Distributing the warnings was another matter. Many on the Keys learned of the forecast by phoning Weather Bureau offices in Key West and Miami; the forecast was then spread by word of mouth. Old-timers had already started preparing for the worst on the thirty-first.

What happened next was an unfortunate but common example of the manifold ways in which people deal with uncertainty in weather forecasts. Experienced residents took no chances and either fled the Keys altogether or did what they could to prepare for the storm. But some of the more recent arrivals, including some of the personnel in charge of the veterans' camps, attached too much certainty to the prediction that the storm would move south of the Keys; they decided to wait. Compounding this problem was the reluctance of the camp administrators to make a definite decision to evacuate, fearing reprimands from their superiors in upstate Florida or Washington should the evacuation prove to have been unnecessary. By the time the football had been passed three or four times back and forth among the various bureaucrats, and an evacuation train finally summoned, it was too late.

The Weather Bureau advisories contributed to the indecision of the veterans' supervisors. The 4 P.M. advisory on September 1 placed the storm "275 miles (450 km) east of Havana and moving west or west-southwest." This was about 160 km (100 mi) east-southeast of where the storm probably was at this time. Evidently, the Bureau realized that its earlier forecasts had the storm moving too fast, and it now overcompensated by moving it too slowly. The wording itself was inadvertently misleading: to all but the most map savvy, the idea of a storm east of Havana conveyed that it was far from the Keys, and its westward movement strongly suggested that it would come nowhere near Florida. Had the Bureau reported the position as being about 480 km southeast of Key West, an alternative description of the same position, residents of the Keys might have been more worried.

As that Sunday wore on, the Bureau compounded its error, reporting at 10 P.M. that the storm was 420 km (260 mi) east of Havana, a movement of only 15 mi in six hours. At this time, the center had passed Andros Island, so the Bureau's position was now in error by about 270 km (170 mi). The error is particularly surprising, given that the surface pressure at Havana was steadily rising, contradicting the idea that a strong hurricane was only 400 km (250 mi) distant and heading straight for it. Moreover, the pressure was now falling rather fast at Miami and Key West, a fact that, taken together with the increasing northeast winds, should have told an experienced forecaster that the storm was well north of the position assigned by the Bureau.

When the hurricane left the Great Bahama Bank in the early morning hours of Labor Day, September 2, it passed over the Gulf Stream, an unusually deep current of very warm water. This was at the time of year when ocean surface temperatures are near their peak. The trajectory of the storm and the time of year were both ideal for a tropical cyclone to attain great intensity.

The 9:30 A.M. advisory on that fateful Labor Day still had the storm centered 320 km (200 mi) east of Havana, an error of at least 480 km (300 mi). Even so, perhaps to cover its bases, the Weather Bureau decided one hour later to hoist hurricane warnings for the Florida Keys. At this point, the veterans' administrators desperately tried but failed to contact their superior in upstate Florida (who was taking a leisurely lunch break followed by a round of golf) to get his permission for a train to evacuate the veterans. A junior official was willing to order the train without such approval, but was overruled by his immediate superior.

The large errors made by the Weather Bureau in estimating the storm's position on Sunday and Monday morning might be partially attributable to the exceptionally small size of the Labor Day

Hurricane. When it passed through the Keys, its eye was estimated to have been only 15 km in diameter, based on the known forward movement of the storm at that time and the duration of the calm at its center. Squally winds extended only about 250 km from the center. By contrast, most hurricane eyes are about 50–60 km in diameter, and squally winds may be found 400–500 km from the center. The eye of a big North Pacific supertyphoon can be 200 km in diameter, with gusty winds extending more than 1,000 km from its center. Thus the entire circulation of the Labor Day Hurricane could fit inside the eye of a Pacific supertyphoon. There is no known relationship between the geometric size of a storm and its intensity, as measured by its maximum wind speeds. Scientists understand why hurricanes cannot get indefinitely big: 1,000 km is an upper bound on the circulation size (a little bigger closer to the equator and a little smaller at higher latitudes), but so far, they have not been able to find a lower limit on hurricane size. There is even a published scientific paper suggesting that the lowly waterspout is really a mini-hurricane, sucking heat out of the ocean on a very small scale.

The tiny size of the storm made it all the more difficult to detect, given the paucity of observations in those days. By 1935 virtually all ships were equipped with radio, so weather reports could be transmitted to the Weather Bureau in time to be useful in locating storms and issuing forecasts. But there is an irony: the better the forecasts, the more ships steer clear of inclement weather, and the fewer are left to report the violent conditions close to storm centers.

Although most Keys residents surmised that a hurricane was approaching that Monday morning, few thought it would arrive as soon as it did, and no one could have imagined just how intense it was. As there were no observations of the storm while it was in the Straits of Florida, I have used a computer model to help estimate the evolution of its wind speed (Figure 19.3) and central pressure (Figure 19.4). There is good agreement between the modeled central pressure and a measurement in the Keys at the time of landfall. Quite a few residents had barometers and recorded the pressure when the storm passed overhead. (One resident of Upper Matecumbe Key noted a reading of 26.00" of mercury (880 mb), the lowest his barometer would register, but in disbelief, he threw the instrument away.) The officially estimated wind speed did not top 160 mph, but it is likely that this will be revised upward when the storm is reviewed in coming years, since the observed central pressure is consistent with a higher wind speed. Also, a post-storm damage survey and engineering analysis estimated wind speeds near 200 mph.

The storm center made landfall near Long Key shortly after 8 P.M. local time (about midnight, Universal Time) on Monday, September 2. Its effects were catastrophic. Owing to the storm's very small size, the winds increased rapidly, taking even the wary by surprise. Gusts of wind estimated to have been in excess of 200 mph obliterated everything from trees to masonry buildings, and a storm surge of between 15 and 20 ft washed over the islands. Almost total destruction occurred in a 40 mi wide swath from Tavernier, just south of Key Largo in the north, to Vaca Key, just north of Marathon. Every single tree and building on Matecumbe Key vanished into the night. The train sent to evacuate the veterans became tangled in the steel cable of a crane that had toppled over the tracks at Windley Key. Shortly after it was freed, around 8 P.M., all the cars except the locomotive were blown off the tracks. People not carried away by the wind were drowned in the quickly rising water; a few survived

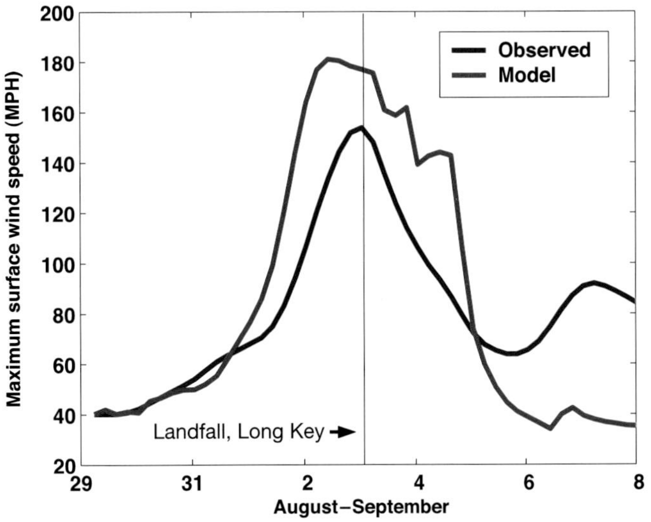

Figure 19.3: Maximum wind speeds in the Labor Day Hurricane of 1935, estimated from observations (blue) and by computer model (red). The vertical line shows the time of landfall in the Keys.

▶

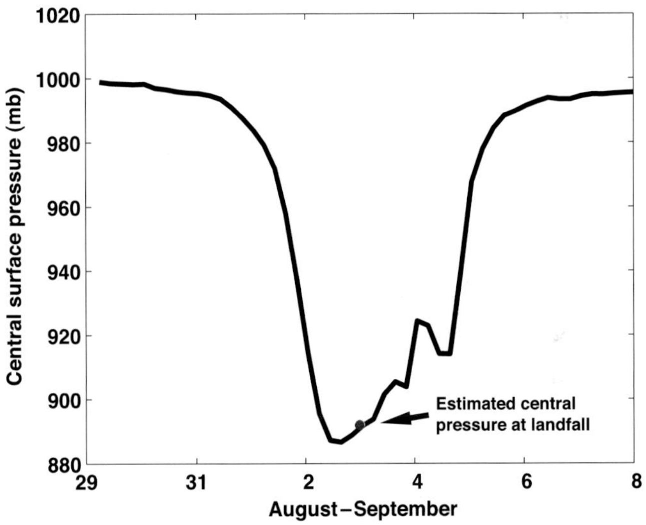

Figure 19.4: Central pressure of the Labor Day Hurricane of 1935 estimated using a computer model. The minimum measured pressure in the Keys is indicated.

▶

after fetching up in the tops of trees. A woman, desperately clinging to her infant son, was blown across 40 mi of open water from Islamorada to Cape Sable, on the Florida mainland. By some miracle, she survived long enough to crawl several hundred feet inland, where the Coast Guard found her body, curled around the corpse of her son. An eight-year-old boy was found the next day clinging with all his strength to a railroad track. Though still alive, he was so traumatized that his hands had to be pried from the track.

When it was through, at least 423 people were dead, including 259 veterans and 164 civilians. The true death toll will never be known, as rescuers, concerned about outbreaks of disease, burned many bodies on the spot; others were never recovered. The majority of those living in the Middle Keys perished, along with a substantial fraction of the Upper Keys population.

The storm dealt a particularly cruel blow to the veterans, who had survived the Great War, the Depression, neglect, and alcoholism, only to be whisked into oblivion by the strongest hurricane ever

Figure 19.5: Veterans Camp 3,
on Lower Matecumbe Key,
before and after the 1935
hurricane.

▶

to strike the United States, while their overseers squabbled over who was going to make the decision to evacuate. "I would rather face machine-gun fire than go through an experience like that again," remarked a surviving vet. On witnessing the carnage, an incensed Ernest Hemingway blamed FDR's government in a passionate article entitled "Who Murdered the Veterans?," published in the radical weekly *The New Masses*. "You're dead now, brother, but who left you there in the hurricane months on the Keys where a thousand men died before you in the hurricane months when they were building the road that's now washed out? Ignorance has never been an excuse for murder or for manslaughter," he wrote, noting that this was not the first time work crews had been killed by hurricanes in the Keys. Charging that the government had been more concerned about completing highways than saving people, he remarked that "there is a known danger to property, but veterans are not property. They are only human beings, unsuccessful human beings, and all they have to lose is their lives."

Figure 19.6: Postcard of the Florida East Coast Railway Key West extension (top) and the wreckage of the veterans rescue train on Windley Key.

▶

Pressure was brought to bear on the Veterans Administration to launch an inquiry into the disaster. Blame was passed around several times before it was finally laid squarely to rest on the one being who could be counted on not to testify in court. The inquiry concluded that the death of the veterans was an "Act of God."

There are those who believe that, thanks to satellites, computers, and modern communications, a Galveston or Labor Day disaster could never happen again. This sense of security is belied by the fact that there are few buildings in the Keys today that would not be submerged by another such storm surge. The only option in the face of an approaching hurricane is to evacuate the entire permanent population of 85,000, plus as many as 50,000 visitors via the mostly two-lane road to the mainland. This would take at least 36 hours. But when Hurricane Andrew bore down on Florida in 1992, 30,000 residents of the Keys flatly refused to leave. Had that storm veered south, the death toll might have been in the thousands.

It is a sad fact that the United States may not have seen its last Galveston.

Storm Tide. Oil painting by American artist Robert Henri, 1903.

The lake's dark breast
Is all unrest,
 It heaves with a sob and a sigh.
Like a tremulous bird,
From its slumber stirred,
 The moon is a-tilt in the sky.
From the silent deep
The waters sweep,
 But faint on the cold white stones,
And the wavelets fly
With a plaintive cry
 O'er the old earth's bare, bleak bones.
And the spray upsprings
On its ghost-white wings,
 And tosses a kiss at the stars;
While a water-sprite,
In sea-pearls dight,

Hums a sea-hymn's solemn bars.
Far out in the night,
On the wavering sight
 I see a dark hull loom;
And its light on high,
Like a Cyclops' eye,
 Shines out through the mist and gloom.
Now the winds well up
From the earth's deep cup,
 And fall on the sea and shore,
And against the pier
The waters rear
 And break with a sullen roar.
Up comes the gale,
And the mist-wrought veil
 Gives way to the lightning's glare,
And the cloud drifts fall,

A sombre pall,
    O'er water, earth, and air.
The storm-king flies,
His whip he plies,
    And bellows down the wind.
The lightning rash
With blinding flash
    Comes pricking on behind.
Rise, waters, rise,
And taunt the skies
    With your swift-flitting form.
Sweep, wild winds, sweep,
And tear the deep

    To atoms in the storm.
And the waters leapt,
And the wild winds swept,
    And blew out the moon in the sky,
And I laughed with glee,
It was joy to me
    As the storm went raging by!

    —Paul Laurence Dunbar (1872–1906),
    "The Rising of the Storm"

20 The Storm Surge

I seek ye vainly, and see in your place
The shadowy tempest that sweeps through space,
A whirling ocean that fills the wall
Of the crystal heaven, and buries all.
And I, cut off from the world, remain
Alone with the terrible hurricane.

—William Cullen Bryant, from "The Hurricane"

One of the most deadly but fascinating phenomena that accompany hurricanes is the storm surge, usually experienced as a sudden rise of sea level near the time of maximum wind speed. This rise of the ocean, which may range from just a few inches to as much as 13 m (42 ft), is historically the most lethal aspect of hurricanes, having killed far more people than wind has.

Storm surges are responsible for the horrifying death tolls—in the hundreds of thousands —of hurricanes striking Bangladesh (Chapter 28), and have been the main cause of death in the two worst hurricanes in U.S. history—the Galveston storm of 1900 (Chapter 13) and the 1928 Okeechobee Hurricane (Chapter 17).

A good if somewhat impersonal description of what it is like to experience a storm surge was provided by Isaac M. Cline, the local forecast official and section director of the Galveston branch of the U.S. Weather Bureau at the time

that city was destroyed by a hurricane in 1900 (see Chapter 13).

Hurricanes affect the elevation of the sea surface in two important ways. First, the very low pressure in the core of the storm literally pulls water up, just like sucking water up a straw. This effect produces one centimeter of sea level rise for each millibar of surface pressure fall; the most intense hurricanes on record have pressure drops of around a hundred millibars, lifting the sea surface by about a meter, or three feet. While this is by no means negligible, it is not the main cause of the storm surge, and its importance is often overstated in popular accounts.

The main driver of storm surges is the stress that the wind exerts on the water. In deep, open ocean, this stress produces complex ocean currents with speeds of a few meters per second. These are associated with slight variations of the sea surface elevation. It is when these currents

Figure 20.1: Wind blowing
across the surface of a
partially filled box piles up
water on its downwind side.

▶

Figure 20.1: Wind blowing across the surface of a partially filled box piles up water on its downwind side.

begin to interact with shallow water and the coast itself that large changes in sea level can occur.

A rough idea of what causes storm surges can be gleaned from a simple thought experiment. Start with an elongated, four-sided box partially filled with water. Blow air across the surface of the water, as shown in Figure 20.1, and wait until the water stops sloshing. Two things happen: the water circulates, flowing with the winds near the surface and returning near the bottom, and its surface tilts upward in the downwind direction. This tilt is needed to balance the friction of the blowing wind: the tendency of gravity to level the water surface balances the force of the wind trying to pile up water on the downwind side of the tank.

If we can neglect friction between the moving water and the tank bottom, we can use a simple formula for the change in elevation of the water at the downwind side of the tank:

$$h = 0.00035 * \frac{V^2 L}{H} , \qquad (20.1)$$

where h is the elevation change in feet, V is the wind speed in miles per hour, L is the length of the box in miles, and H is the *average* water depth, in feet. This illustrates several interesting properties: The water rises quickly with wind

speed, increasing as its square, and also rises with the distance (*fetch*) over which the wind is blowing. Also, the smaller the average water depth, the greater the rise.

For example, a 100 mph wind blowing over 100 miles of water that is 30 ft deep would produce an 11 ft increase in the water level on the downwind side of the box.

Although this simple picture captures some of the important physics of storm surges, nature is more complicated. In the first instance, water trying to pile up against a straight beach can flow off to the sides, unlike in our box, and this reduces the surge. But surges can be large in narrow bays, from which water has nowhere to escape.

Another way to increase the storm surge above the level given by equation (20.1) is by resonance. A simple example of resonance is a child on a swing. Like a pendulum, the swing has a natural period of oscillation; by pumping the swing in tune with its natural period, it can be made to go quite high.

Suppose we abruptly switch off the fan in our simple thought experiment. The water in the box would slosh back and forth with a frequency determined by the length of the box and the average depth of the water, gradually slowed to rest by friction. To make a really big storm surge in our box, we could switch off the fan, wait for the water to slosh back to the upwind side of the box, and then, just as it starts to move back toward the downwind side, switch it on again. By continuing to switch the fan off and on like this, we could create very big surges, just like pumping a swing.

This type of resonance is hard to achieve in the ocean, because unlike our box, it is open on at least one end. But it is not hard to do in an enclosed body of water like a lake. When a hurricane moves over a shallow lake, such as Florida's Lake Okeechobee, the wind blows first from one direction, piling water up at the downwind side of the lake. But as the eye passes, the wind switches and blows from the opposite direction. If the timing is right, the resulting surge can be significantly greater than if the wind had just blown from a single direction. A resonant sloshing of lakes or bays is called a *seiche*.

Several other types of ocean oscillation can partially resonate with a moving hurricane's wind and pressure fields and create storm surges bigger than predicted by equation (20.1). Some of these oscillations can produce large surges even when a hurricane is moving parallel to the coast, rather than at oblique angles to it. To forecast storm surges, we must use sophisticated computer models that take into account the distribution, speed, and direction of the evolving winds; the movement of the storm; the bathymetry of the seafloor; and the shape of the coastline, including its bays, inlets, peninsulas, and islands. These models are successful in forecasting storm surges, given good forecasts of the hurricanes themselves.

A few rules of thumb have emerged from experience with these models, as well as basic physical reasoning. For storms approaching the coast head on, the surge will be larger if

- the storm has a lower central pressure;
- the onshore component of wind speed is larger;
- the eye is larger (although there is not much effect for eyes larger than about 30 mi);
- the storm is approaching the coast faster;
- the seafloor shoals gradually;
- or the coastline is part of a bay that is as narrow as or narrower than the eye of the storm.

For storms approaching the coast obliquely or traveling nearly parallel to the coast, the surge will be larger if

- the storm is moving along the coast with the coastline to its right, in the Northern hemisphere, or left, in the Southern Hemisphere;
- the storm is moving fast, if the coastline is to its right (Northern Hemisphere), or slowly, if the coastline is to its left (Northern Hemisphere);
- or the storm's eye is larger.

One other complicating factor is the normal ebb and flood of the astronomical tides. In some places, such as Florida, the normal tide range is barely a foot, so it hardly matters whether an intense storm strikes at high or low tide. In New England, on the other hand, normal tides range from a few feet along the Connecticut shore to more than 15 ft at Eastport, Maine. Even an intense hurricane striking eastern Maine at low tide may not produce a surge that extends above the normal high-water mark, whereas a strike at high tide can be devastating.

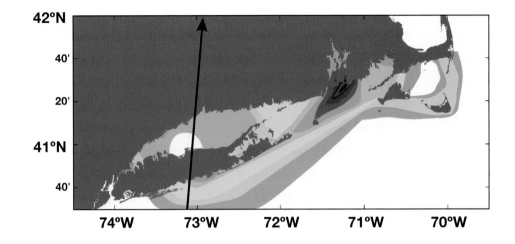

Figure 20.2: The maximum storm-surge height during the New England Hurricane of 1938. Contour interval is 0.3 m, with height ranging from 1.8 m to 5.1 m. The storm center moved along the path shown by the arrow.

▶

An example of the variability of storm surge heights is shown in Figure 20.2 for the New England Hurricane of 1938, which was moving exceptionally fast when it made landfall (Chapter 21). The largest sea level rises were in Narragansett Bay; in general the largest rises were in bays to the right of the track. Providence, Rhode Island, at the head of Narragansett Bay, is frequently flooded during storms. At the peak of the 1938 hurricane, much of its downtown was underwater.

The rise and fall of the sea can be very complex, as demonstrated by the evolution of the sea level height at Atlantic City, New Jersey, during the passage of a severe storm (not a hurricane) offshore (Figure 20.3). The normal astronomical tide is also shown. There were three main surges: one just after high tide and another two several hours later, almost at the time of low tide.

Sometimes water levels rise and fall dramatically several times within a few hours, owing

Figure 20.3: Storm surges recorded at Atlantic City, New Jersey, during the passage of a storm offshore. The solid curve shows the normal swing of the tides, the dashed curve shows the observed sea level, and the dotted curve shows the part of the sea level rise attributable to the storm surges.

▶

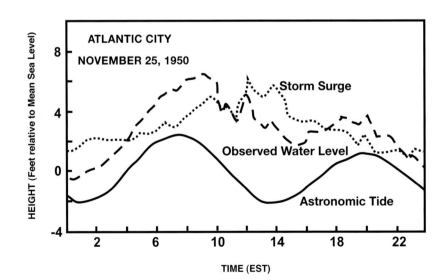

Figure 20.4: The Great Gale of 1815, *a contemporary painting by John Russell Bartlett, depicts flooding at Providence, Rhode Island, from a storm surge during the hurricane.*

▲

to resonance between the storm surge and oscillations that can occur in shallow water. The second rise, which can be larger than the first, is sometimes referred to as a *resurgence.* Most of the damage that Hurricane Carol inflicted on Cape Cod, Massachusetts, in 1954 was a result of a resurgence.

Sea level often begins to rise a few hours, and sometimes days, before the storm surge itself arrives. These sea level increases, called *forerunners,* may be caused by large storm-generated swells that pile water up against the coast.

Atop the storm surge ride enormous waves driven by the hurricane's winds. These act as battering rams, greatly increasing the destruc-

tion caused by the storm surge flooding. The combination of storm surge, huge waves, howling winds, and flying debris makes survival extremely problematic for those unfortunate enough to be caught in low-lying coastal terrain during a hurricane.

The ancient Mayans learned by experience to build their cities inland, away from the reach of hurricanes and their storm surges. Yet despite centuries of hurricane disasters, our society continues to disregard collective experience and invite future tragedy by building more and more structures in surge-prone coastal regions. Short memory is usually cited as the main reason for this irrational behavior; after all, very few of

Figure 20.5: Storm surge of Hurricane Carol floods a Rhode Island yacht club in 1954.

▶

those at risk have experienced a major hurricane. But policies in force in the United States create strong incentives to build on floodplains. Storm surge damage is not covered under private insurance policies but by the Federal Flood Insurance Program, whose government-set rates do not reflect the great variation of risk from one place to another. Moreover, the private insurance industry is heavily regulated by individual states, under the oversight of insurance commissioners who, in some states, are elected. Rates are set not by insurance companies, but by the state, which is under popular pressure to make insurance available to everyone who wants it and to keep rates reasonable, even for those who choose to assume large risks. The result is a government that strongly subsidizes risky behavior. Given the long history of entanglement of the insurance industry with the state, and the enormous investments individuals and organizations have made in coastal property, it is unlikely that this system will change anytime soon, as we march blindly on to the next hurricane disaster.

The Hurricane. Oil painting by American artist Louis Monza, 1956.

Lord of the winds! I feel thee nigh,
I know thy breath in the burning sky!
And I wait, with a thrill in every vein,
For the coming of the hurricane!
And lo! on the wing of the heavy gales,
Through the boundless arch of heaven he sails;
Silent, and slow, and terribly strong,
The mighty shadow is borne along,
Like the dark eternity to come;
While the world below, dismayed and dumb,
Through the calm of the thick hot atmosphere
Looks up at its gloomy folds with fear.
They darken fast—and the golden blaze
Of the sun is quenched in the lurid haze,
And he sends through the shade a funeral ray—
A glare that is neither night nor day,
A beam that touches, with hues of death,

The clouds above and the earth beneath.
To its covert glides the silent bird,
While the hurricane's distant voice is heard,
Uplifted among the mountains round,
And the forests hear and answer the sound.
He is come! he is come! do ye not behold
His ample robes on the wind unrolled?
Giant of air! we bid thee hail!—
How his gray skirts toss in the whirling gale;
How his huge and writhing arms are bent,
To clasp the zone of the firmament,
And fold, at length, in their dark embrace,
From mountain to mountain the visible space.
Darker—still darker! the whirlwinds bear
The dust of the plains to the middle air:
And hark to the crashing, long and loud,
Of the chariot of God in the thunder-cloud!

You may trace its path by the flashes that start

From the rapid wheels where'er they dart,

As the fire-bolts leap to the world below,

And flood the skies with a lurid glow.

What roar is that?—'tis the rain that breaks,

In torrents away from the airy lakes,

Heavily poured on the shuddering ground,

And shedding a nameless horror round,

Ah! well-known woods, and mountains, and skies,

With the very clouds!—ye are lost to my eyes.

I seek ye vainly, and see in your place

The shadowy tempest that sweeps through space,

A whirling ocean that fills the wall

Of the crystal heaven, and buries all.

And I, cut off from the world, remain

Alone with the terrible hurricane.

> —William Cullen Bryant (1794–1878),
> "The Hurricane"

21   The Great New England Hurricane of 1938

*I sometimes feel that we have had a preview
of the end of the world.*

—Anne Moore, survivor

On September 4, 1938—almost exactly a year before Hitler's Germany invaded Poland and precipitated World War II—a French weather observer at the Saharan station of Bilma Oasis noted a gentle wind shift, signaling the passage of another of the easterly waves that move across Africa and out over the Atlantic with great regularity in summer. Little did he know that this wave was destined to spawn the greatest natural disaster in New England history, leaving 680 people dead amidst terrible wreckage. Like many natural disasters, the Great New England Hurricane was a lethal combination of meteorological chance and human error.

The next definitive observation of the system was made by the SS *Algegrete*, encountering a full-blown hurricane northeast of Puerto Rico on September 16. This drew the attention of the U.S. Weather Bureau office in Jacksonville, Florida, which then had jurisdiction for all tropical weather south of Cape Hatteras. From the sixteenth until midday on the nineteenth, the strengthening storm headed straight toward southeast Florida (Figure 21.1), prompting the Weather Bureau to post a hurricane warning there. Landfall was expected less than a day later, and a storm-wary state prepared for the worst.

But by the morning of September 20, it became clear that the storm was slowing and turning toward the north, sparing Floridians still reeling from the violent hurricanes of 1926, 1928, and 1935. Warnings were dropped because the storm was expected to continue curving off to the northeast, as such storms usually do. But it did not behave like most hurricanes. To the great misfortune of New Englanders, the September 1938 storm was poised to begin a deadly dance with a meteorological partner from the north.

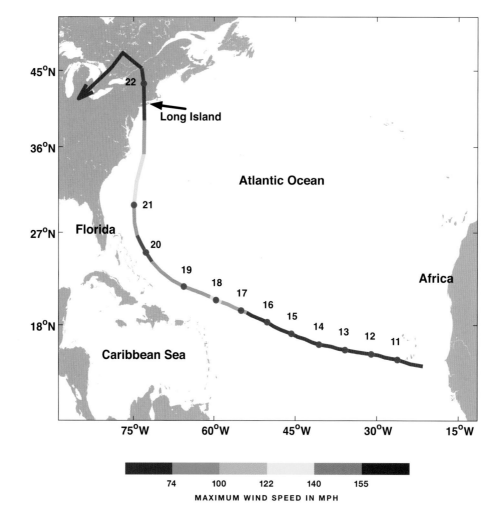

Though there were few upper-air observations in 1938, there were enough to show that the Great Hurricane was introduced to its high-latitude consort on September 20 (Figure 21.2). At 4 A.M., as the hurricane skirted the eastern Bahamas, a strong upper-air trough, extending southward from the Great Lakes, swept eastward as though to keep a rendezvous with its tropical date.

Such troughs are everyday features of weather outside the Tropics. To their east, southwesterly winds bring warm, humid air up from the south, while west of the trough, cold air plunges southward from Canada. As they migrate eastward, the warm, southerly current ascends, and as it does so, the water vapor within it condenses into clouds and precipitation. Conversely, the cold air west of the trough descends and dries, bringing bright, clear skies. These troughs, unlike hurricanes, derive their energy from the contrast between cold air to the north and warm air to the south. The system of troughs and ridges moving eastward across the middle latitudes is responsible for most of the alteration between fair and foul weather found almost everywhere outside the Tropics. The trough over the Great Lakes on the morning of September 20 was one of these systems, perhaps a little stronger than average, but very definitely in the wrong place at the wrong time.

Around midnight on the twentieth, the Great Hurricane began to be swept off her feet by the trough, whose increasing southwesterly winds accelerated his partner northward. Thus began the

Figure 21.2: Map of pressure and winds 10,000 ft (3 km) above the surface at 5 A.M. on September 20, 1938, constructed from balloon observations. Lines of equal pressure are shown by black curves, with L and H denoting low and high pressure. Red arrows show wind direction.

▶

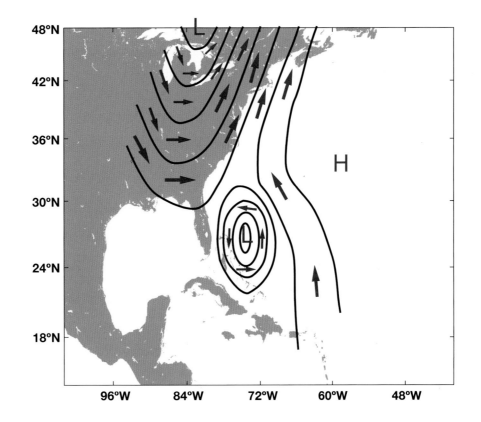

Figure 21.3: Map of pressure and winds 10,000 ft (3 km) above the surface at 5 A.M. on September 21, 1938, constructed from balloon observations.

▶

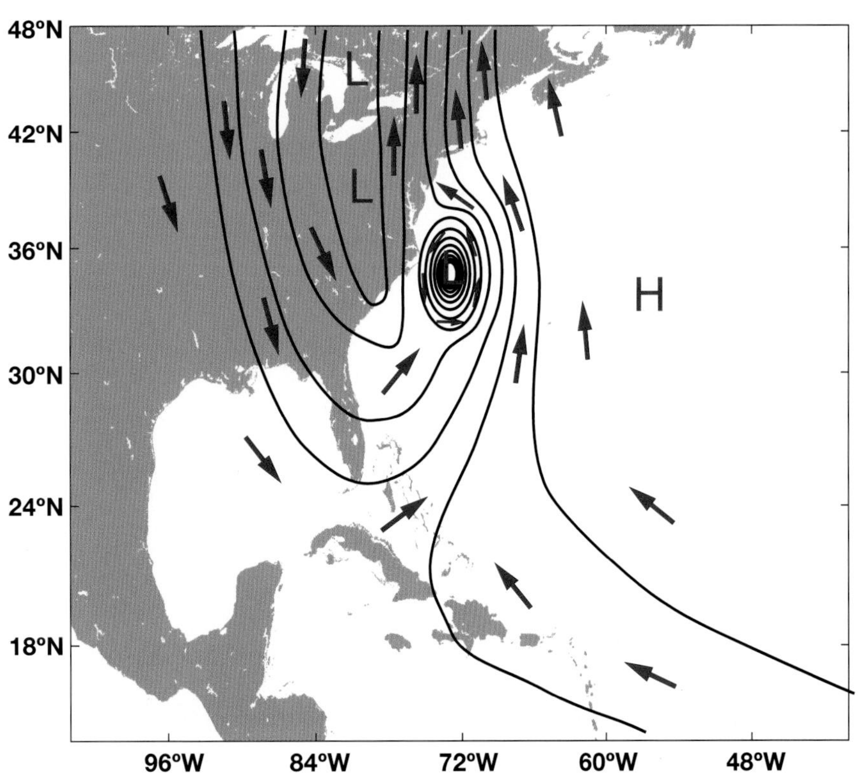

great atmospheric duet that meteorologists call extratropical transition (Chapter 16). Normally, hurricanes weaken rapidly as they pass over the much colder water north of the Gulf Stream, but the Great Hurricane had a dance partner that continually fed her energy and kept her winds strong. Just as important, the strong winds ahead of the trough swept the hurricane northward so fast that barely 12 hours elapsed before it left the warm, supportive waters of the Gulf Stream and slammed headlong into Long Island. With well-matched dance partners, each dancer supports the other: while the upper-air trough fed energy into the hurricane, the counterclockwise circulation around the storm drew even more cold air southward from Canada, thereby strengthening the trough aloft. So far south was the cold air drawn in this case that, by early on September 21, the winds ahead of it had switched to south (Figure 21.3), swiftly carrying the Great Hurricane to its rendezvous with New England.

It happened so fast that the U.S. Weather Bureau was caught off guard. Early on the twenty-first, as the storm came abreast Cape Hatteras, Jacksonville followed established procedure and turned forecast responsibility over to the Washington, D.C., branch of the Weather Bureau. The first ominous sign that something was amiss with the forecast was an observation transmitted shortly after midnight on the twenty-first by the Cunard liner *Corinthia*, which, trying desperately to clear the treacherous waters off Cape Hatteras, reported a pressure of 943 mb (27.85")—extremely low for a storm so far north, and probably not even as low as the actual central pressure of the storm. Other ship observations indicated that the storm was well west of where the Weather Bureau thought it was at that time, but there were not enough observations to reveal the terrific forward speed the hurricane had achieved.

Charles Pierce, an alert 28-year-old junior forecaster, analyzed what few observations were available and surmised that the hurricane was accelerating northward. He argued that hurricane warnings should be posted for Long Island and New England, but was overruled by his superiors. Just 12 hours before the storm hit Long Island, there were no storm warnings posted north of Atlantic City; in fact, no actual hurricane warnings were ever posted north of Florida.

On the morning of Wednesday, September 21, residents of New York's Long Island and of New England awoke to leaden skies and depressing newspaper headlines. British prime minister Neville Chamberlain was on his way to Munich to negotiate the partitioning of Czechoslovakia with Adolf Hitler. France and England had acceded to Hitler's demands, and war now seemed inevitable. In the midst of such distressing news appeared a rosy editorial in the *New York Times* noting that "the cyclone that happily spared our Southern coast is not an extraordinary occurrence" and boasting that "if New York and the rest of the world have been so well informed about the cyclone it is because of an admirably organized meteorological service." Buried in the same newspaper was the forecast for New England: "Rain, probably heavy today and tomorrow, cooler."

And so it came to pass that the Great Hurricane of 1938 struck Long Island and New England with no warning. The last major hurricane to affect the region was in 1815, and it had since become common wisdom that hurricanes avoided New England. The summer had been extraordinarily wet: many towns and cities reported the third wettest June in their history. The week before the storm was particularly rainy, with several inches recorded in many places. But, oddly, September 21 dawned mostly clear and warm, with a gentle southeast breeze. Fishermen put to sea and children went out to play.

By around 10 A.M., the sky had clouded over in Long Island and southern New England, and the wind had picked up considerably. Quite a few people noticed that the sky had assumed an odd, yellowish tint. Still, with no warnings posted, few took precautions. Now moving northward at 60 mph, the center of the storm slammed ashore around 3 P.M. near Moriches Inlet, about ten miles west of Westhampton Beach, Long Island, accompanied by a ten-foot storm surge riding atop a spring tide already one foot above normal. Some witnesses recall seeing what they at first took to be a fog bank, only later realizing with horror that they were looking at a wall of water. The force of the impact was registered on seismograms as far away as Sitka, Alaska. Near and to the east of the center, winds increased to over 100 mph, while in New York City, winds of over 80 mph were recorded, with a gust of 120 mph at the top of the Empire State Building. The lowest pressure reading, 945 mb (27.94"), was observed at Bellport Coast Guard Station, about 15 mi west of Westhampton Beach. The combination of wind, storm surge, and tremendous waves obliterated most shorefront property—the towns of Kismet, Fair Harbor, Saltaire, and Cherry Grove were almost completely destroyed. The calm of the storm's rather large eye—about 50 mi in diameter—was felt from Brentwood in the west to Mattituck in the east. Many residents, with little or no experience of hurricanes, ventured out to inspect the damage, only to be caught out when the storm's fierce southern eyewall overtook them.

The hurricane roared northward across Long Island Sound, passing near Milford, Connecticut, and half an hour later over Hartford, which recorded a minimum pressure of 950 mb (28.04"). The strongest winds were felt to the east of the center, where the forward motion of the storm added to the vortical motion around its center. Along the south coast of Connecticut, winds in excess of 100 mph were observed in several places, from New London to Fall River, and were estimated to have peaked at 120 mph at the Weather Bureau station on Fishers Island, near New London, just after the weather tower blew down. As far east as Providence, sustained winds of 100 mph and gusts to 125 mph were recorded. Continuing up the Connecticut River valley, the storm passed just west of Northfield, Vermont, and nearly over Mount Whiteface, New York, where a ten-minute calm interceded hurricane-force winds from the northeast and, after, from the southwest.

The storm moved so rapidly over New England that it simply did not have time to lose much force, taking only a few hours to pass from Long Island to Canada. Friction greatly reduced the average wind speeds near the land surface, but only a short distance aloft winds well over 100 mph were recorded from New York City almost to Boston. The large wind shear in the first 1,000 ft or so no doubt produced terrific wind gusts at the surface, especially over hilly terrain. The observatory atop Blue Hill, just south of Boston, recorded sustained winds of 120 mph for five minutes, shortly after 6 P.M. By then, the main anemometer had disintegrated, and the staff meteorologist had to rely on a standard three-cup anemometer. This instrument was not designed to record gusts, but based on subsequent examination of the record, wind gusts as high as 186 mph were estimated to have occurred. On top of New Hampshire's Mount Washington, which four years earlier had set the world's surface wind speed record of 231 mph, sustained winds of 118 mph were augmented by gusts to 163 mph, destroying much of the cog railway that served the observatory.

The storm surge, well over 10 ft in places, inundated most of Long Island's barrier beaches, from Fire Island inlet to Southampton, carving a new inlet to Shinnecock Bay. Even bigger surges flooded the central Connecticut coastline, ranging between 14 and 18 ft west of New London and from 18 to 25 ft from New London to Cape Cod (see Figure 20.2). A surge of some 12 to 15 ft raced up Narragansett Bay, destroying almost everything along the waterfront and ultimately flooding downtown Providence to a depth of 20 ft. The prevailing sense of unreality was heightened by the sight of headlights shining through the water from thousands of submerged automobiles, and the sounds of wailing sirens and shorting auto horns. In Rhode Island, many smaller oceanside towns, like Misquamicut and Napatree Point, were completely obliterated by the storm surge.

The effects of the surge were truly horrifying. Had the storm struck a few weeks earlier, many of the summer beachfront cottages would have been occupied, and the death toll might have been far higher. As it was, few of those remaining survived. Of the 179 houses on Westhampton Beach, 153 completely vanished.

There were many tales of death and unlikely survival. The Moore family of Napatree Point watched in terror as their beachfront neighbors were carried off their front porch by the storm surge. When their own house began to disintegrate, they congregated in the attic, which floated free when the house collapsed into the surging ocean. Catherine and Jeffrey Moore, their four children, an aunt, a maid, and a handyman clung to what had been the attic floor, with one wall still attached as it sailed across Little Narragansett Bay and fetched up in Connecticut. All survived.

As the great storm was approaching Long Island, the express train *Bostonian* left New York City on its regular run to Boston. With no warning that a storm was on its way, the chief engineer did what he could to keep the train moving through driving rain and increasing wind. After seeing trees and telegraph poles begin to topple, he determined to reach the relative safety of Stonington Station. But as the train crossed a narrow, exposed causeway stretching two thousand feet across the waters of Long Island Sound, it was signaled to stop by a red light indicating another train ahead. Just after permission to proceed had been secured, the storm surge struck, washing out the road bedding under the last three cars. The train's crew uncoupled these cars after moving all the passengers forward, but by then the train was so entangled in wreckage it could not move. Finally, all the passengers and crew were herded into the forwardmost car, which, pulled by the still-encumbered engine, finally managed to move forward. But their ordeal was not over. First a boat and then an entire house drifted onto the tracks, but in each case the train simply plowed through the obstacles, ultimately reaching Stonington.

During a hurricane's extratropical transition, the distribution of rainfall shifts markedly to the west of the storm track, departing from the far more symmetrical distribution of tropical hurricanes. True to form, the Great New England Hurricane brought heavy rains to central and western New England, along and west of the storm track. Though heavy, this rainfall was relatively short-lived because the storm was moving so rapidly. Nevertheless, this proved the straw that broke the camel's back.

For four days, beginning on the previous Sunday, September 18, New England had experienced almost unprecedented heavy rain. By the morning of the day the storm struck, the Connecticut, Thames, and Merrimack rivers were all in flood, and the water levels continued to rise. The Chicopee River was already 5½ ft above its previous record level, set in 1936, while the Millers River at Orange and the Aschuelot at Hinsdale were, respectively, 3½ and 2 ft above their previous record levels. The Deerfield River at Shelburne Falls also exceeded its previous record level. In some places, more than ten inches of rain had fallen. Much of the rain fell in cloudbursts of short duration, contributing to the severity of the flooding. It is unusual for New England rivers to flood in September, and these floods, by themselves, constituted a disaster. It was bad luck that they were followed immediately by the strongest hurricane ever recorded in the region.

To the unprecedented precipitation of the last four days, the Great Hurricane added 4 to 6 more inches over the central and western hills of New England, pushing the four-day total over 17 inches in some spots. Consequently, many rivers and their tributaries rose to extraordinary levels, wreaking havoc. The town of Ware, Massachusetts, was particularly hard hit (Figure 21.4). In Hartford, the Connecticut River rose 19.4 ft above flood stage. The water-soaked ground made it easy for the wind to topple trees, and many an elm, much beloved in New England towns, succumbed to the onslaught. It has been estimated that in just a few hours, as many as a quarter billion trees were uprooted in New England, with some 2.6 billion board feet of timber affected. Sea salt, thrown up with the copious spray accompanying the storm, blew far inland, killing vegetation as far as 20 mi from the coast.

Figure 21.4: Flooding took out this bridge at Ware, Massachusetts.

►

Figure 21.5: The lighthouse tender Tulip *blown ashore across railroad tracks in New London, Connecticut.*

▶

The Great New England Hurricane and floods took the lives of 680 people and seriously injured another 1,750. More than 93,000 families suffered major property losses, including nearly 7,000 summer dwellings, 2,000 other dwellings, 2,600 boats, and 2,300 barns. About 26,000 automobiles were destroyed. Somewhere between 500,000 and 750,000 chickens were killed, as well as 1,675 head of livestock. The storm downed nearly 20,000 mi of electric and telephone lines, cutting service to more than 80 percent of electricity customers and 30 percent of all telephones in New England. Railroad service between New York and Boston was halted for 7 to 14 days while 10,000 workers filled in 1,000 washouts, replaced almost 100 bridges, and removed countless obstructions from the tracks, including 30 boats and one very large steamer (Figure 21.5). All told, the storm did more than $300 million worth of damage in 1938 dollars, or about $4 billion 1998 dollars, making it the sixth-costliest natural disaster in U.S. history. Had today's infrastructure existed then, it is estimated that the storm would have racked up almost $18 billion in losses.

*Figure 21.6: Downtown
Hartford, Connecticut, after
the '38 hurricane.*

▶

*Figure 21.7: The Long Island
shore after the '38 hurricane.*

▶

The Eye of the Hurricane.
Oil painting by Dutch artist
Engel Hoogerheyden,
ca. 1795.

As one that in a silver vision floats
Obedient to the sweep of odorous winds
Upon resplendent clouds, so rapidly
Along the dark and ruffled waters fled
The straining boat.—A whirlwind swept it on,
With fierce gusts and precipitating force,
Through the white ridges of the chafèd sea.
The waves arose. Higher and higher still
Their fierce necks writhed beneath the tempest's scourge
Like serpents struggling in a vulture's grasp.
Calm and rejoicing in the fearful war
Of wave ruining on wave, and blast on blast
Descending, and black flood on whirlpool driven
With dark obliterating course, he sate:
As if their genii were the ministers
Appointed to conduct him to the light

Of those belovèd eyes, the Poet sate
Holding the steady helm. Evening came on,
The beams of sunset hung their rainbow hues
High 'mid the shifting domes of sheeted spray
That canopied his path o'er the waste deep

—Percy Bysshe Shelley (1792–1822),
from "Alastor: Or, the Spirit of Solitude"

22   **Waves**

Emerald chasms parting widely,
In between the alpine waves;
Yawning like the mouth of Hades
Downward into Neptune's cave.

—Friedrich Schiller, from "Hero and Leander"

No aspect of a hurricane is more frightening than its enormous waves. Many a ship has succumbed to them, and breaking waves are responsible for much of the damage ashore, even in storms that never make landfall.

Waves higher than 60 ft have been recorded in hurricanes, but worse than these colossal heights is the turmoil of the seas near the eye, where the wind direction shifts very quickly. Over the centuries, various strategies have been developed to help ships cope with large waves, but these are useless when waves are coming from many different directions, adding to one another and creating a state of maritime chaos. Adding to the sailor's misery is the horrifying prospect of a rogue wave, a solitary monster arising without warning from the general chaos and quickly overwhelming anything afloat, from a dory to a supertanker (Figure 22.1).

WAVE PHYSICS

A gentle breeze blowing across the water excites myriad ripples, shimmering in the sunlight. These are called capillary waves, and they owe their existence to surface tension, the same force that holds water droplets together. Deflect the water surface upward, and surface tension will force it back down; this *restoring force* is the primal ingredient of all oscillations.

A stronger breeze will create bigger waves whose restoring force is gravity. The basic physics of such waves is simple: gravity accelerates patches of elevated water downward and adjacent depressions upward. But the factors that determine the actual sea state are complex.

Waves are described by their speed and direction of travel, their steepness, and whether or not they are breaking. The sea surface usually undulates with a spectrum of waves having differ-

Figure 22.1: A rogue wave looms astern of a merchant ship in the Bay of Biscay. Although this photo was not taken in a hurricane, it shows the kind of sea one might encounter in such storms.

▶

ent heights, wavelengths, and speed and direction of travel.

The speed at which wave crests move depends on several factors. Waves in deep[15] water move at a speed given by

$$c_{deep} \approx \sqrt{\frac{gL}{2\pi}},$$

where g is the acceleration of gravity and L is the distance between wave crests (i.e. the *wavelength*). The longer the wave, the faster it moves. For example, a wave whose wavelength is 100 ft will move at about 22 ft/s (15 mph), given that g is 32 ft/s^2. Waves in shallow water[16] move at a speed given by

$$c_{shallow} \approx \sqrt{gh},$$

where h is the ocean depth.[17] Unlike its deep-water counterpart, the speed of a wave in shallow water does not depend on its wavelength. In water that is 100 ft deep, for example, such waves move at 57 ft/s (39 mph).

Deep-water ocean waves, whose propagation speed depends on their wavelength, are said to be *dispersive*. Suppose a stretch of ocean contains waves of varying wavelength, all moving in the same direction. Because the longer waves move faster than the shorter ones, the waves will sort themselves by wavelength, with the longest moving out in front and the shortest falling to the rear. Dispersive waves have the curious property that their energy moves at a speed different from that of individual wave crests.

---

[15] Deep compared to the distance between two wave crests.

[16] Shallow compared to the distance between two wave crests.

[17] A general expression for c, valid for all ranges of ocean depth, is $c \approx \sqrt{\frac{gL}{2\pi} \, tanh \, \frac{2\pi h}{L}}$, where *tanh* is the hyperbolic tangent function. This reduces to the shallow and deep expressions described earlier in the limits of large and small L.

Figure 22.2: Wave packets modulate the size of individual waves within them. In the case of deep-water waves, the packets move at half the speed of individual wave crests.

▶

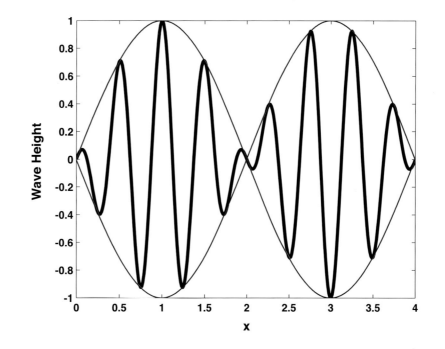

Try following the crests of ocean waves as they move toward the beach, focusing, say, on a particularly large one. This becomes a frustrating experience as the large wave shrinks and may become so small as to be untraceable. But just as it does so, the next wave begins to grow until it becomes the largest one in sight. This sequence repeats itself until, finally, the largest wave enters shallow water and breaks.

In this example, the wave heights slowly change from one place to another. These height variations take the form of a kind of superwave that modulates the individual waves. This superwave is called a *wave packet* and is illustrated in Figure 22.2. The thick curve represents individual waves that are moving, say, from left to right. As they move, they have to stay within the envelope of the superwave or wave packet, shown by the thin curve. So as they approach x=2, where the envelope disappears, their height vanishes, only to reemerge at larger values of x. (Here x represents distance.) There is one complication: the wave packet is also moving.

The energy of waves moves at the packet speed, not the speed of individual wave crests. This packet speed is called the *group velocity*. In the deep ocean, the group velocity is one half the speed individual wave crests move. (The energy of nondispersive waves, like shallow-water waves, moves at the same speed as individual wave crests.)

The concepts of wave packets and group velocity help us understand how ocean waves grow, move, and decay. Waves are produced by wind: the energy input to waves increases with the cube of the wind speed. The longer the wind acts on wave packets, the larger the waves, until a point is reached when they lose energy to dissipation or transfer to other waves at the same rate they are gaining energy from the wind. When waves interact with each other, energy is usually transferred "upscale," from shorter to longer waves, and when waves break, their energy is turned into heat.

Now consider ocean wave packets on four sides of a Northern Hemisphere storm moving

Figure 22.3: Wave packets produced by a Northern Hemisphere hurricane moving toward the northwest. For simplicity, we consider wave packets in four quadrants of the storm.

▶

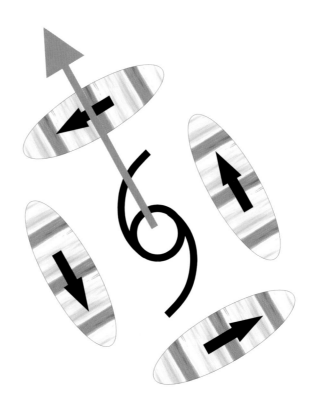

toward the northwest, as shown in Figure 22.3. Northeast of the storm's center, the southeasterly winds generate a wave train moving northwest, in the same direction as the storm. The energy input to the waves is largest at the radius of maximum wind speed. Since the wave packets are moving along with the storm, they will be exposed to strong winds for a long time, and the waves within them can thus become very large. Conversely, on the southwest side of the storm, wave packets are moving toward the southeast and are not exposed to strong winds for very long, so they have little opportunity to grow large. Northwest and southeast of the center, wave packets are exposed to high winds for an intermediate duration.

Of all the waves in a hurricane, the biggest are those moving at about one-third the wind speed. The group velocity of these waves is thus about one-sixth the wind speed (in deep water). The most prodigious seas therefore arise in storms moving at about one-sixth of their maximum wind speed (e.g., a storm with 100 mph winds moving at 17 mph); in this case, the wave packets containing the largest waves are moving right along with the storm, and the waves have plenty of time to grow.

Although wind energy is deposited preferentially into wave packets moving at around one-sixth the wind speed, collisions between waves of different wavelengths transfer energy to longer waves, which move faster. So wave energy generated on the right-hand side of the storm will tend to disperse forward to the right-front quadrant.

Mariners have known for centuries to stay clear of the right-front quadrant of hurricanes (left-front in the Southern Hemisphere), where the strongest winds and highest waves are found. Today, we can measure wave heights over large regions of the storm using an instrument called a scanning radar altimeter. This device is flown on an airplane and sends pulses of radiation down to

Figure 22.4: Wave heights (in meters) in Hurricane Bonnie of 1998, as determined from an airborne scanning radar altimeter. The axes show distance east and north of Bonnie's center.

▶

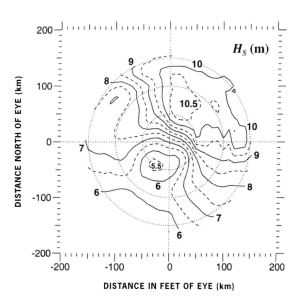

Figure 22.5: Height and direction of propagation of waves in Hurricane Bonnie of 1998, as detected with a scanning radar altimeter. The black barbs point in the direction of wave propagation, and their width is proportional to the wave height. The thin barbs show the direction of the low-level winds.

▶

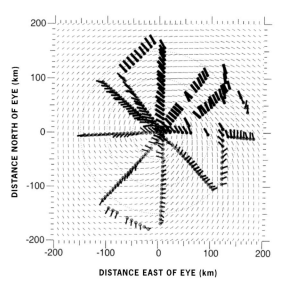

the sea surface, sweeping back and forth across the flight track. By measuring the time it takes the pulses to reflect off the water and return to the radar, one can determine the altitude of the sea surface, and thus measure the heights of the waves. Figure 22.4 shows the wave heights around Hurricane Bonnie in August 1998. Bonnie was moving slightly west of north at this time. As expected, the biggest waves are in the right-front quadrant.

This radar can also detect the wavelength and direction of wave propagation, as shown in Figure 22.5. Note that the waves are often moving at oblique angles to the wind direction. A more detailed analysis of the radar data shows that there are some places where two or more wave trains are crossing paths, creating a chaotic sea state. Analyses like these are giving researchers new insights into the nature of hurricane waves.

Wave interaction shunts energy to longer and faster waves that outrun the storm. Often the first sign of an approaching hurricane is the very long swell that has traveled far out ahead of the storm. As the storm draws closer, the wavelength of the most prominent waves slowly decreases, as does the time interval between the passage of wave crests (i.e., the wave period). In the days before aircraft reconnaissance, meteorologists carefully timed the period of waves breaking on the beach as a means of detecting an approaching storm.

The sound of long period swell breaking on beaches and headlands is one of the distinctive traits of approaching hurricanes. The energy of these advance foot runners is often large enough that the sound of their breaking carries many miles inland, alerting the wise to approaching danger.

This brings us to the question of what happens to deep-water waves as they approach coastlines. As the bottom shoals to a depth comparable to their wavelength, they begin to slow down: both their propagation speed and their group velocity decrease. The first to feel the effects of the bottom are the longest waves.

As wave packets approach the shore, their front side slows down first, causing the rear part of the packet to catch up to it, like rush hour commuters slowing at a traffic jam. The wave packet energy piles up into an ever-decreasing volume, causing the waves themselves to amplify and eventually break. This energy, which may have been generated thousands of miles offshore, is dissipated in shifting sands and eroding rock.

Waves often refract as they approach the shore. Those approaching at an angle gradually bend around to become nearly parallel to the beach by the time they break. Waves encountering a peninsula bend around it, focusing their energy on its tip; conversely, waves entering a bay refract radially outward, sparing the beaches at the head of the bay.

Hurricane waves pounding ashore, often riding atop a storm surge, release enormous quantities of energy accumulated over days or weeks at sea. So powerful is the pounding that it can sometimes be detected by seismographs thousands of miles away. Feared by mariners and beachfront residents alike, waves often administer the coup de grâce to ships and buildings battered by the storm's furious winds.

Figure 22.6: Wind-driven waves pour over a seawall during a hurricane near Miami, Florida.

▶

Figure 22.7: Huge waves batter Old Lyme, Connecticut, during Hurricane Carol.

Figure 22.8: Typhoon-generated waves strike a breakwater in Japan.

A tramp Steamer
In a Gale.

▲

Tramp Steamer in a Gale.
Drawing by Joe Spedding, in
1933 at age 14, Lepe, UK.

At its setting the sun had a diminished diameter and an expiring brown, rayless glow, as if millions of centuries elapsing since the morning had brought it near its end. A dense bank of cloud became visible to the northward; it had a sinister dark olive tint, and lay low and motionless upon the sea, resembling a solid obstacle in the path of the ship. She went floundering towards it like an exhausted creature driven to its death. The coppery twilight retired slowly, and the darkness brought out overhead a swarm of unsteady, big stars, that, as if blown upon, flickered exceedingly and seemed to hang very near the earth. …A faint burst of lightning quivered all around, as if flashed into a cavern—into a black and secret chamber of the sea, with a floor of foaming crests.

It unveiled for a sinister, fluttering moment a ragged mass of clouds hanging low, the lurch of the long outlines of the ship, the black figures of men caught on the bridge, heads forward, as if petrified in the act of butting. The darkness palpitated down upon all of this, and then the real thing came at last.

It was something formidable and swift, like the sudden smashing of a vial of wrath. It seemed to explode all round the ship with an overpowering concussion and rush of great waters, as if an immense dam had been blown up to windward. In an instant the men lost touch of each other. This is the disintegrating power of a great wind: It isolates one from one's kind. An earthquake, a landslip, an avalanche, overtake a man incidentally, as it were—without passion. A furious gale attacks him like a personal enemy, tries to grasp his limbs, fastens upon his mind, seeks to rout his very spirit out of him….

After the whisper of their shouts, their ordinary tones, so distinct, rang out very loud to their ears in the amazing stillness of the air. It seemed to them they were talking in a dark and echoing vault.

Through a jagged aperature in the dome of clouds the light of a few stars fell upon the black sea, rising and falling confusedly. Sometimes the head of a watery cone would topple on board and mingle with the rolling fury of foam on the swamped deck, and the *Nan-Shan* wallowed heavily at the bottom of a circular cistern of clouds. This ring of dense vapours, gyrating madly round the calm of the centre, encompassed the ship like a motionless and unbroken wall of an aspect inconceivably sinister. Within, the sea, as if agitated by an internal commotion, leaped in peaked mounds that jostled each other, slapping heavily against her sides; and a low moaning sound, the infinite plaint of the storm's fury, came from beyond the limits of the menacing calm. Captain MacWhirr remained silent, and Jukes' ready ear caught suddenly the faint, long-drawn roar of some immense wave rushing unseen under that thick blackness, which made the appalling boundary of his vision….

He watched her, battered and solitary, labouring heavily in a wild scene of mountainous black waters lit by the gleams of distant worlds. She moved slowly, breathing into the still core of the hurricane the excess of her strength in a white cloud of steam—and the deep-toned vibration of the escape was like the defiant trumpeting of a living creature of the sea impatient for the renewal of the contest. It ceased suddenly. The still air moaned. Above Jukes' head a few stars shone into the pit of black vapours. The inky edge of the cloud-disk frowned upon the ship under the patch of glittering sky. The stars too seemed to look at her intently, as if for the last time, and the cluster of their splendour sat like a diadem on a lowering brow.

—Joseph Conrad (1857–1924), *Typhoon*

▶

The Big Blow (A destroyer rides out a typhoon in the Sea of Japan, ca. 1950). Drawing by U.S. Navy combat artist Herbert C. Hahn.

23   Bull Halsey's Typhoons

I am more afraid of the West Indian Hurricane than I am of the entire Spanish Navy.

   —President William McKinley, 1898

Preoccupation with the job at hand, or a desire not to disturb the skipper, should never result in disregard of a rapidly falling barometer.

   —Fleet Admiral Chester W. Nimitz, 1945

When in 1274 Kublai Khan launched his first great naval invasion of Japan, he had no knowledge of great storms at sea and no means to detect them. By the time Admiral William F. "Bull" Halsey steered the U.S. Third Fleet directly into the cores of two typhoons in December 1944 and June 1945, conflicting objectives and overreliance on sparse, inconsistent data had replaced meteorological ignorance as the main cause of two great maritime disasters.

On the evening of December 17, 1944, Halsey was trying desperately to decide what course of action to take, balancing his military objective of supporting General MacArthur's invasion of the Philippines with conflicting and uncertain reports of a typhoon nearby.

Two days earlier, forecasters at Fleet Weather Central in Pearl Harbor, Hawaii, examined sparse observations from ships, islands, and aircraft and concluded that a tropical disturbance of some kind was making up in the western Pacific. They located the storm slightly to the south of its likely actual position at the time, but incorrectly believed it was tracking to the north-northwest when, in fact, it was headed only slightly north of west.

The fleet's weather officer, Commander George F. Kosco, was a seasoned forecaster, or "aerologist" as they were known in those days. He had considerable experience with hurricanes and, looking at the same island observations as Fleet Weather Central, came to roughly the same conclusion about the disturbance's location. He also observed that a cold front was progressing toward the fleet and would overtake them the next day. He forecast that the tropical disturbance, which he then believed to be weak, would become involved in the frontal zone and race northeastward. Even by today's standards, this was by no means an unreasonable expectation.

At midday on the seventeenth, Kosco estimated that the fleet, which was already experiencing weather rough enough to cancel a scheduled refueling operation, was in the middle of a weak cold frontal zone and that a tropical storm (not yet thought to be a typhoon) was about 400 mi to the southeast.

In the early morning hours of the same day, a Navy pilot discovered a tropical disturbance with 70 mph winds about 250 mi southeast of the Third Fleet's position. But the cumbersome secret coding practices of the day seriously delayed the transmission of that information to Halsey's staff, and the visual observations by the pilot failed to identify the disturbance as the full-fledged typhoon it had likely become. Moreover, the reported position of the storm was about 100 mi south of its actual location, which was much closer to the fleet. It was an unfortunate and costly fact that the U.S. Navy, which assigned low priority to weather reconnaissance, declined to provide the Third Fleet with a dedicated weather aircraft.

Meanwhile, Fleet Central, choosing to disregard the aircraft observation, merely extrapolated the position of the storm based on its analysis of its location and movement two days earlier. By 3 P.M. on the seventeenth, the analyzed location was a remarkable 600 mi northeast of the storm's likely position at that time.

Based on these conflicting reports as well as his own direct observations of the worsening conditions, Kosco recommended that the fleet steam northwestward, at right angles to the cold front. He still believed that the storm was weak and would race northeastward once it encountered the front. Halsey agreed with the recommendation and, at 1:30 P.M., ordered the fleet to proceed northwestward, even though this course would place the fleet in danger of Japanese kamikaze attacks. Halsey, as it were, had to decide which "divine wind" was worse. Little did he or his weather officers suspect that the supposedly weak and distant tropical storm was a raging typhoon that would not be slowed or turned by a weak cold front, or that it was only 140 mi to the southeast.

And yet the signs were there for those willing to read them. By noon, the barometer was falling rapidly on Halsey's flagship, the *New Jersey*, with an increasing wind from the east-northeast. The falling barometer was inconsistent with a frontal zone retreating to the southeast, and an ominous and increasing swell from the east should have been an additional clue to the real state of affairs. The Third Fleet's weather officers were no doubt misled at this time by both the aircraft observation, which located the storm too far south, and Fleet Weather Central's analysis, which had the storm far to the northeast and retreating. Both reports located the storm too far east. To a trained meteorologist, logic and basic training would dictate that a violent storm was approaching from the southeast and that the safest course for the fleet would be to turn south and maneuver to the storm's less dangerous left side, where the winds and seas are generally less violent. But Kosco was worried about the aircraft report, which, had it been accurate, would have suggested that a southerly course would take the fleet right into the core of the storm. Finally, acting mostly on his own direct observations of the wind, seas, and falling barometer, Kosco recommended that the fleet head south.

After much discussion with his weather team, Halsey decided at 3:50 P.M. to steer the fleet south toward a new rendezvous point, 185 mi south of the original meeting area. In the event that the tropical storm continued to move toward the northwest, as by now they correctly supposed it was doing, this course would place them in the less dangerous semicircle of the storm.

At this point, there is some confusion about the decisions that were made. After some ships reported that they were not making enough progress southward, another rendezvous point was arranged roughly halfway between the first and second locations. But by now, some of the fleet had steamed south of the latitude of this new rendezvous. Halsey's weather officer advised him to continue south, but Halsey made the fateful decision to turn the fleet back north. It is easy to find fault with Halsey, and to accuse him of ignoring the advice of his weather officer in favor of a plan that better suited his military objective. But Halsey had been receiving conflicting advice up until that time, was concerned about the fleet's being found by the enemy, and was not convinced of the existence of a genuine typhoon.

It is not clear why the fleet had not by this time detected the typhoon by radar. It is true that radar had only been around a few years, and its use detecting weather phenomena was not yet well developed. (The first radar image of a hurricane eye had been taken only three months earlier at the U.S. Naval Air Station at Lakehurst, New Jersey.) Yet published photographs of the radar aboard the Third Fleet's USS *Wasp* show clear evidence of a typhoon eye a short distance southeast of the fleet.

The weather continued to deteriorate during the night of December 17–18. Earlier on the seventeenth, the fleet was battered by winds so fierce that the pilots of two aircraft, returning from a sortie, were told not to try to land on the pitching aircraft carrier and were instead directed to bail out of their planes, to be fetched from the sea by a destroyer. By the early morning hours of the eighteenth, the weather had become so bad that Halsey convened his officers to consider what to do next. After some discussion, it was decided to turn the fleet south again, and the order went out at 4:57 A.M.

There is little controversy about what happened next. Encountering frightful seas and extreme winds, the ships scattered in many directions as they fought for survival.

Trouble developed early aboard the light carrier *Monterey*. Captain Stuart Ingersoll ordered all the ship's aircraft secured to the deck by half-inch cable. But at 0910 he radioed that all planes in his hanger deck were on fire. Breaking loose in the tumult below decks, these fuel-laden loose cannons ricocheted off each other and the hanger walls, throwing off showers of sparks and igniting fuel. The flames were sucked down ventilation shafts, setting fires on the decks below. (The location of the ventilation ducts on the inside of the hanger deck were one of many shortcuts taken in the rush to build ships at the outbreak of the war.)

As the captain hove to, sailors fought bravely to put out the fires. One sailor, caught by a 100 mph gust, was blown straight off the deck into the ocean. Seeing this, a shipmate unrolled a long fire hose into the sea. By shear good luck, the sailor surfaced right next to the hose and, grabbing it, hauled himself back aboard. Others were not as fortunate: many suffered from burns and smoke inhalation. Among those nearly swept overboard was future U.S. president Gerald Ford, who was then serving as assistant navigator, athletic officer, and antiaircraft battery officer onboard the *Monterey*. Finally, the fire was brought under control, and the *Monterey* limped back to Ulithi.

Within the fleet, matters progressed from bad to worse. Destroyers such as the *Aylwin* were already top-heavy from the addition of heavy guns, radar, and other equipment high up on the ship. In the heavy seas and winds of the typhoon, they began to roll violently, especially when trapped in the

Figure 23.1: The light carrier
Langley *lists sharply to star-*
board as she is buffeted by
winds of the first of Halsey's
typhoons.

▶

wave troughs. The *Aylwin* rolled 70° to each side, hanging precipitously on the verge of capsizing each time, and taking on tons of seawater through her blower intakes. With loss of engines and rudder control, she drifted helplessly before the onslaught. Two men, who had taken to the deck to escape the extreme heat of the engine room, were swept cleanly overboard, never to be heard from again.

Blinded by blowing spray and rain, unable to steer freely if at all, and having in many cases lost radar and radio communication, the captains of the Third Fleet's ships added collision to their list of fears. To make matters worse, tanks, guns, bombs, and 300 lb depth charges broke loose and went careening over the decks. Numerous fires broke out and had to be extinguished.

When he received the order to come to a heading of 140°, Commander James A. Marks of the destroyer *Hull* attempted to comply. Just as the ship was coming on course, she was overtaken by a 120 mph blast and set skidding across the ocean, completely out of control. She began a death roll, at first dipping 50° from side to side, then 70°, until finally she was forced all the way over on her side, at which point she was swamped by the giant waves. What crew could do so scrambled onto the upward-facing side and abandoned ship, clinging to floating debris and a few rafts that had broken free. Some 20 survivors would later be rescued, but 202 of the *Hull*'s crew died that day.

A similar fate befell the destroyer *Monaghan*. Commander Floyd B. Garrett, trapped in mountainous seas, made a futile attempt to take on water ballast, but the pumps sucked air and seized. The ship was so battered by the winds and seas that her seams began to leak, eventually flooding the engine room and causing the loss of all power and steering. As in the case of the *Hull*, the disabled ship commenced a death roll, resulting in her capsize. Only 13 men managed to climb aboard a raft, and of these, all but 6 were eventually lost to the sea, to sharks, or to dehydration. The remaining 6 were picked up three days later, all that survived of *Monaghan*'s men.

The commander of the destroyer *Spence* elected to discharge the seawater ballast she had taken aboard in expectation of heavy weather, so she could take on fuel. When the refueling operation was abandoned owing to the heavy seas, the *Spence* was left floating high above the surface, exposed to the full power of the wind and seas. Caught by the wind, the ship was blown 75° over to port and held there for ten seconds, losing all power. Then she fell over on her side. Of the sailors onboard, roughly 200 went down with the ship; many more drowned or were eaten by sharks. A few made it onto floating debris, only to be blown out of sight forever. Most of the few that survived did so by clinging to a floating net; a few were also rescued from a raft.

All told, only 74 of the combined complement of 831 men onboard the *Hull*, *Monaghan*, and *Spence* survived. Including casualties on other ships, some 790 men lost their lives in the typhoon. Seven ships were seriously damaged; some limped into port with missing or crushed bows. Some 186 airplanes were destroyed, having been burnt, jettisoned, or blown overboard.

In hindsight, Halsey's Third Fleet might be regarded as a victim of a weather forecasting system in rapid transition. Hundreds of years of experience had long before led to procedures for detecting and dealing with violent cyclonic storms at sea, based solely on observations of clouds, winds, waves, and, eventually, barometers. By these means, one can make good inferences about the bearing and direction of movement of intense cyclones within a few hundred miles of the ship, and take appropriate actions to get out of harm's way or, at least, maneuver into the least dangerous sectors of the storm. But beginning early in the twentieth century, telegraph and later radio made it possible to map the distributions of pressure, wind, temperature, and precipitation. Such analyses clearly revealed the existence and movement of cyclones, anticyclones, and fronts. This new era of synoptic meteorology represented a great leap in understanding and predicting weather and took much of the guesswork out of forecasting out to a day or so. Commander Kosco and his weather officers were well trained in this new science.

But synoptic meteorology relies on weather observations that are dense enough in space and frequent enough in time to delineate weather systems such as fronts and hurricanes. Such observations as were available to the Third Fleet were grossly inadequate to the task confronting them in December 1944. Had Commander Kosco and his weather officers fallen back on the tried-and-true canon of an earlier day, relying solely on their own observations, they might have concluded 6 to 12 hours earlier than they did that an intense cyclone was bearing down on them from the southeast, and persuaded Halsey to take the fleet south, out of the path of the storm. Such armchair speculations as these, however, must be tempered by the fact that, at the time, the forecasters faced making very tough decisions based on conflicting evidence, in the stress of war at sea.

Less than six months later, Halsey's men and ships once again confronted a typhoon. On June 3, 1945, a reconstituted Third Fleet was positioned off Okinawa when ships and search planes reported a storm making up from the south. The fleet's meteorologist, after studying the reports, recommended that the fleet remain near Okinawa, which did not appear to be in the path of the storm.

But Halsey argued that should the storm change course, he did not want to be forced westward into shallow water with no maneuvering room, and within range of Japanese aircraft in China. Halsey ordered his fleet to set a course toward the southeast, and at the same time sent the amphibious command ship *Ancon* toward the presumed position of the storm to report weather conditions. On the evening of June 4, *Ancon* sighted a typhoon on her radar, but once again this critical piece of information did not reach Halsey expeditiously, not arriving until 1 A.M. on June 5. Had it arrived on time, Halsey might have asked the *Ancon* to track the storm; instead, her captain, no doubt acting prudently, ordered the ship out of harm's way.

Figure 23.2: Halsey's second typhoon took the bow clear off the Pittsburgh *(top) and warped the flight deck of the* Hornet *as though it were made of cardboard (bottom).*

▶

On receiving the typhoon report, Halsey immediately convened his weather staff and asked if their present heading of 110° would take the ship out of the course of the storm. The meteorologists thought not. Seeking to get his ships out of the dangerous right-hand quadrant of the storm approaching from the southwest, Halsey ordered a change of course to 300°. Once again, this turned out to be directly into the path of the very small but intense typhoon. By the time the storm was spotted on the fleet's radars, it was too late.

This time, it was not the destroyers that took the worst beating, but the larger cruisers and carriers. Encountering seas of 75 ft and winds to 135 mph, the fleet was seriously damaged. The cruiser *Pittsburgh* had 104 ft of her bow wrenched off; this was later rounded up by a tugboat and brought back to port. All four carriers were damaged. The forward flight decks of the *Hornet* and *Bennington* collapsed, and the *San Jacinto*'s hull buckled. Some 76 aircraft were lost. Six men were killed and four more seriously injured.

Nature did far more damage to the Third Fleet then the enemy ever did. Writing in his autobiography, Admiral Halsey described his impressions thus:

> No one who has not been through a typhoon can conceive its fury. The 70-foot seas smash you from all sides. The rain and the scud are blinding; they drive you flat-out, until you can't tell the ocean from the air. At broad noon I couldn't see the bow of my ship, 350 feet from the bridge. The *New Jersey* once was hit by a 5-inch shell without my feeling the impact; the *Missouri*, her sister, had a kamikaze crash on her main deck and repaired the only damage with a paintbrush; yet this typhoon tossed our enormous ship as if she were a canoe. Our chairs, tables and all loose gear had to be double-lashed; we ourselves were buffeted from one bulkhead to another; we could not hear our own voices above the uproar.

The enormous loss of life and damage from Halsey's typhoons prompted the Navy to establish a dedicated typhoon reconnaissance operation in the western Pacific, which continued until 1987; they also created the Joint Typhoon Warning Center, originally built in Guam and now located in Hawaii. With a dedicated team of forecasters, and satellite surveillance that virtually guarantees that no typhoon will go undetected, it seems unlikely that future ship commanders will ever face the paralyzing lack of information that so confounded Bull Halsey. Even so, Admiral Chester W. Nimitz's advice to future sailors remains as valuable as ever:

> The time for taking all measures for a ship's safety is while still able to do so. Nothing is more dangerous than for a seaman to be grudging in taking precautions lest they turn out to have been unnecessary. Safety at sea for a thousand years has depended on exactly the opposite philosophy.

Sudden Rain Shower on Ohashi Bridge. Woodcut by Japanese artist Ando Hiroshige, 1857.

R ain-obliterated
The river, some roofs
A bridge without a shore

—Matsuo Bashō (b. 1624)

24   Ratin

I don't say it rained, because it was like another deluge.

—Christopher Columbus

Hurricanes are usually thought of primarily as wind storms. Virtually all metaphorical use of the word *hurricane* in literary works evokes violent wind. Yet some of the worst tropical cyclone catastrophes are caused not by winds but by torrential rain. Curiously, some of the most devastating floods are produced by tropical cyclones of sub-hurricane strength.

A recent and especially tragic example is that of Hurricane Mitch of 1998, the deadliest Atlantic hurricane since the Great Hurricane of 1780 (Chapter 11). Floods produced by Mitch killed more than 11,000 people in Central America, and the president of Honduras declared that Mitch destroyed 50 years of progress in that country.

Mitch developed as a rare Category 5 hurricane in the central Caribbean, with sustained winds in excess of 175 mph. It drifted slowly southward over Honduras and neighboring countries, producing rainfall rates of 1 to 2 ft per day and precipitation totals of as much as 75 in. The resulting floods and mud slides virtually destroyed the entire infrastructure of Honduras and devastated parts of Nicaragua, Guatemala, Belize, and El Salvador. Whole villages together with their inhabitants were swept away by the resulting flash floods and mudflows.

MEASURING HURRICANE RAIN

Before the advent of meteorological radar, there were few reliable measurements of rainfall at sea in hurricanes. On land, rain gauges are virtually useless during high winds, with rain blowing almost horizontally. Everything changed when the first radar image of a hurricane was made in 1944. Although it is not altogether simple to estimate actual rates of precipitation, radar estimates of rain in the hurricane core are much better than rain gauge data, with all of its problems in high winds. And, of course, there are few rain gauges at sea.

Figure 24.1: Hurricane Mitch, as photographed from a geostationary satellite, at 12:45 UT on October 26, 1998.

▶

Figure 24.2: Radar reflectivity measured from a hurricane reconnaissance aircraft flying in the eye of Hurricane Floyd of 1999. Map is 360 km (225 mi) square. The radar reflectivity measures roughly how much rain, snow, and hail is in the air.

▶

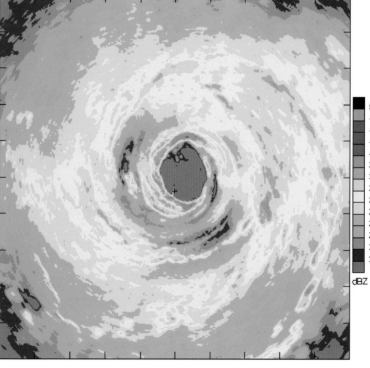

Figure 24.3: Radar reflectivity in Hurricane Gilbert of 1988, made from a reconnaissance aircraft flying at 2.6 km altitude. The black lines show the aircraft track, and the black barbs show the direction and speed of the wind measured by the aircraft. Each short barb is worth 5 kts, the longer barbs 10 kts, and the triangles 50 kts. The map is 240 km square.

▶

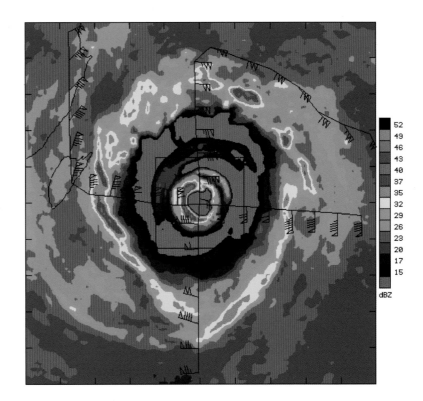

Hurricanes exhibit a variety of features on radar displays, like the one shown in Figure 24.2. Around a clear central eye can be seen the ring of heavy eyewall rainfall. Outside of that is a region of reduced rain surrounded by another ring of high rainfall. Floyd was displaying a fairly common feature of intense hurricanes: a double eyewall structure, with inner and outer eyewalls and reduced rainfall in between. These concentric eyewalls evolve through a characteristic sequence, with an outer eyewall developing outside an existing eyewall, contracting inward, and choking the inner eyewall like a boa constrictor. The entire evolution may take a day or so, and sometimes hurricanes may undergo two or more eyewall cycles. A particularly dramatic example of a double eyewall is that of Hurricane Gilbert in 1988 (Figure 24.3), the most intense storm ever recorded in the Atlantic. There is no radar return whatsoever between the two eyewalls, a region sometimes referred to as the "moat."

Beyond the outer eyewall in Floyd (Figure 24.2), to the east and south of the center, are two prominent spiral rain bands. Features like these were first noticed in the early days of radar and can also be seen in satellite images, such as the one of Mitch in Figure 24.1. The rainfall rates in spiral bands are sometimes as large as those of the eyewall. Theories to explain spiral bands abound, but none is generally accepted by researchers today.

In 1997, meteorologists acquired a new tool for measuring precipitation at sea. That year, NASA launched the Tropical Rainfall Measuring Mission (TRMM), which for the first time put a meteorological radar into orbit. Although its spatial resolution is not as good as ground-based radars, TRMM can see precipitation everywhere, including over the oceans. The rainfall rate it estimated in an Indian Ocean tropical cyclone is shown in Figure 24.4, superimposed on a conventional infrared image of the storm.

Figure 24.4: Radar reflectivity in an Arabian Sea tropical cyclone, measured from space in NASA's Tropical Rainfall Measurement Mission (TRMM). The colors show the radar reflectivity, while the white shading represents a conventional infrared image of the storm's clouds.

▶

The eye, the eyewall, and an outer rain band are clearly visible in the TRMM radar reflectivity.

HURRICANE RAINFALL PHYSICS

There are four factors that determine how much rain will fall in a given place: the amount of water vapor in the air, how fast the air ascends, and the vertical extent and duration of the updraft.

The updraft speed and the amount of moisture in the air determine how fast water vapor condenses into cloud. As air ascends, its pressure drops, since pressure decreases with altitude. According to the first law of thermodynamics, falling pressure causes falling temperature, at a rate of about 1°C for each hundred meters the air ascends. The amount of water vapor air can hold declines with falling temperature, and so the air's capacity for water vapor eventually falls below its actual water vapor content. When this happens, the vapor begins to condense into the tiny droplets we see as cloud. Condensation heats the air, so the rising air does not cool as fast as before.

One might guess that the rate of precipitation is just determined by how moist the air rising through cloud base is, and how fast it rises. Alas, it is not that simple. Most of the time, clouds are surrounded by clear, dry air, which mixes into them and reevaporates some of the condensed water. In small cumulus clouds, all of the water

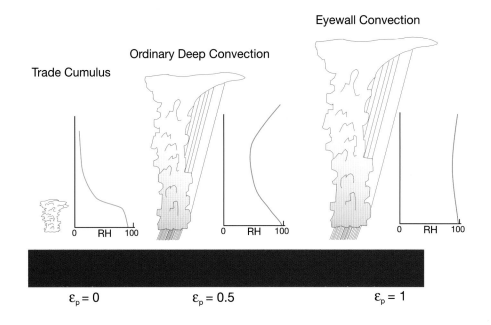

Figure 24.5: The precipitation efficiency of three types of atmospheric convection. "Trade" cumulus clouds (left) are shallow and ascend into dry air; all the water droplets eventually evaporate and so the precipitation efficiency is zero. In ordinary thunderstorms (center), roughly half the water vapor that rises through cloud base eventually reaches the surface as rain. In the humid eyewalls of hurricanes, almost all the water vapor that rises through cloud base falls to the surface as rain. In each figure, the graphs show typical vertical profiles of relative humidity.

▶

that condenses eventually reevaporates, so there is no rain, even though moist air is ascending into the clouds. Larger clouds do produce rain, but some of it evaporates on its way to the surface.

A deep cloud produces more rain than a shallow one, even if its updraft speed is the same, because the air ascends to a greater altitude and so loses more water. But there is an added benefit: Precipitation falling from the top of the cloud sweeps up some of the tiny cloud droplets it collides with on the way down.

Meteorologists define a quantity called the *precipitation efficiency* as the fraction of water vapor ascending through cloud base that ultimately makes it to the surface as precipitation. The precipitation efficiency of shallow cumulus clouds is zero since they do not produce rain at the surface. Generally, the deeper the cloud and the more humid its environment, the greater its precipitation efficiency. The eyewall of an intense hurricane is quite exceptional: it is nearly saturated with water vapor everywhere, so there is very little reevaporation of condensed water. The

eyewall clouds are also very deep, extending all the way through the troposphere and even into the lower stratosphere. Consequently, the precipitation efficiency in the core of a strong hurricane can be almost 100 percent. The different factors affecting the precipitation efficiency of shallow cumulus clouds, ordinary thunderstorms, and hurricane eyewalls are illustrated in Figure 24.5.

Since eyewalls are always tall and their precipitation efficiency is usually large, the main factor controlling the rainfall rate in eyewalls is the updraft speed, which is in turn controlled mainly by two factors, at least while the storm is over the ocean.

First, when the storm is intensifying, a deep layer of moist air must flow into the core of the storm, where it turns and flows upward. The more rapidly the storm is intensifying, the stronger the ascent in the core.

The second factor is friction between the storm's winds and the underlying surface. Even when a hurricane has stopped intensifying, friction causes air near the surface to flow inward

Figure 24.6: Airflow in a circular vortex without (a) and with (b) friction. The circles are curves of equal pressure (isobars), with lowest pressure at the vortex center. The "pressure gradient force" accelerates air toward lower pressure. Without friction (a), this force is exactly balanced by the centrifugal force of the air going around the vortex. Friction slows this airflow (b), reducing the centrifugal force and allowing air to spiral in towards the storm center.

▶

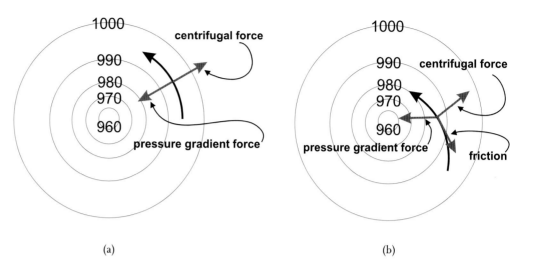

(a)   (b)

toward the eyewall, where, again, it must turn upward. To see why, first consider a circular vortex on which no friction acts (Figure 24.6a). Air pressure decreases inward toward the center of vortices. This *pressure gradient* pushes air toward the storm center. But since the air is spinning around the vortex center, another force, the *centrifugal force*, accelerates air outward, away from the center. In a steady, frictionless vortex, these forces exactly balance, and the air flows around the storm center at constant speed. This is in some ways similar to the orbits of planets, in which gravity balances centrifugal force.

Near the surface, friction acts to slow the air, producing a three-way force balance, as illustrated in Figure 24.3b. The pressure gradient force is unchanged, pushing air toward the storm center, but the centrifugal force is acting in a slightly different direction, both outward and forward, because the airflow is now curving around a different center. A third force, friction, acts to slow down the air and is directed opposite to the air motion. In steady flow, these three forces exactly balance, so the speed of the airflow is constant. The air, rather than flowing in circles around the center, spirals inward. As it converges

on the center, it has nowhere to go but up. Thus friction acting on a vortical flow forces upward motion in the vortex core. This frictional inflow, combined at times with the upflow associated with storm intensification, produces the ring of intense rainfall surrounding the eye, as can be seen in satellite photos such as that of Hurricane Mitch in Figure 24.1 and, especially, in radar images such as the one of Hurricane Floyd shown in Figure 24.2.

Over the ocean, then, rainfall is a function of both the wind speed, which determines how much frictional inflow will occur, and the rate of intensification of the storm, which is related to the strength of the nonfrictional inflow. How much rain actually falls at any fixed point in space also depends on how long it rains. The duration of the rain is proportional to the breadth of the storm divided by the speed at which it is moving. So big, slow-moving storms produce the most rainfall at any given point.

The average hurricane at sea precipitates about one trillion gallons of rainwater per day, about triple the rate of freshwater consumption in the United States in 1995.

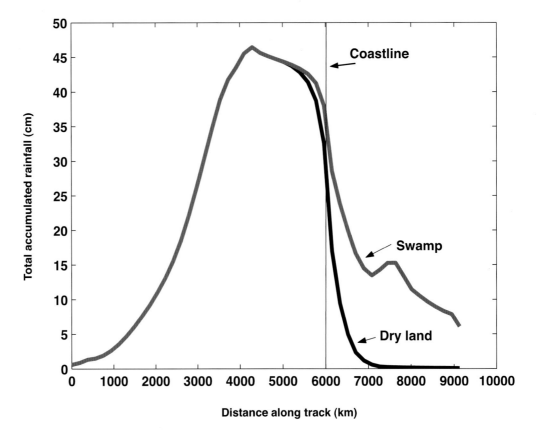

Figure 24.7: Total amount of rainfall accumulated at points 24 km to the right of the track of a Northern Hemisphere hurricane simulated by a computer model, displayed as a function of the distance along the storm's track. The storm is assumed to be moving at 15 mph. On the left side, the storm intensifies over the ocean and its rainfall increases. After traveling 6,000 km, the storm makes landfall. Over dry land, the rain quickly diminishes as the storm moves inland, while over a swamp it lasts somewhat longer.

▶

RAINFALL OVER LAND

When a hurricane makes landfall, the physics that control rainfall change dramatically. On the one hand, friction between the winds and the surface generally increases, thereby increasing frictional inflow. This favors more rain. On the other hand, the vortex rapidly loses strength, and the surface does not supply as much evaporation, particularly if the ground is dry. Usually, these last two factors dominate, and rainfall diminishes quickly after landfall. But if the land is wet, some evaporation occurs, enhancing the amount of water vapor flowing into the storm. Also, standing water, such as a swamp, can add enough heat to the storm to substantially slow the storm's decay (Chapter 16). Figure 24.7 shows the total accumulated rainfall 24 km (15 mi) to the right of the track of a hurricane moving 15 mph in a

straight line perpendicular to the coast, as simulated by a simple computer model.

As the hurricane intensifies offshore, its rainfall increases. About 2,000 km (1,200 mi) from the coast, the storm settles down to a nearly steady intensity, with slightly diminished rainfall. Starting several hundred kilometers (about 100 miles) offshore, rainfall totals begin to decrease, even though wind speeds (not shown) remain very high until the storm center crosses the coast.

After crossing onto dry land, rainfall diminishes rapidly, from about 25 cm (10") right at the coast to a still respectable 5 cm (2") 500 km inland. But if the hurricane passes over swamp instead of dry land, the rainfall decays much more slowly. At the coast itself, the total rainfall is about 32 cm (12.5"), compared to 25 cm in the case of dry land. But only 200 km inland, there is

Figure 24.8: Map showing
accumulated rainfall, in mil-
limeters, from Hurricane
Mitch in Nicaragua.

▶

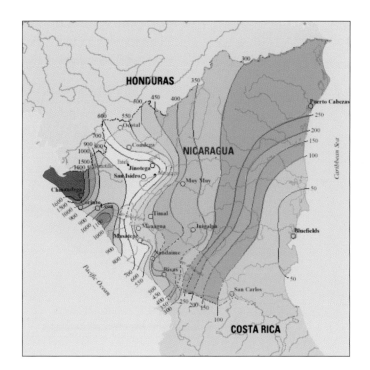

almost 25 cm of rain over the swamp, compared to 10 cm over dry land.

If the storm is moving half as fast, roughly twice as much rain falls on a given location. The speed at which the storm moves is a critical factor in determining how much rain falls.

After a hurricane makes landfall, it still possesses a deep, very moist inner core. Rainfall diminishes largely because the updrafts weaken when the storm winds die down. But several factors may serve to make the moist air rise again.

Mountains are among the most important of these factors. Although surface winds decrease rapidly, wind speeds a kilometer or two above the surface can remain high. If the storm encounters mountains, this air is forced to ascend along the windward slopes, and very heavy rainfall can result. Some of the worst flooding disasters in history have happened when hurricanes or their remnants pass over mountains.

In Hurricane Mitch, many factors conspired to produce exceptionally heavy rain. First,

Mitch began as an unusually intense storm over very warm ocean water, so it was already producing very heavy rain when it made landfall on the north coast of Honduras. Second, Mitch stalled once it crossed the coastline. And finally, the interior of Central America is very mountainous, with peaks in Honduras as high as 2,800 m.

Figure 24.8 shows a map of total rainfall in Nicaragua as a result of Mitch, which passed to its north and west. Most of the high terrain in Nicaragua is in the northwest part of the country. The counterclockwise flow around Mitch brought in additional water vapor from the Pacific, and its strong northwesterly winds were forced to rise over the mountains, producing more than a meter of rain.

The heaviest rainfalls in history have occurred on high, mountainous islands that either receive glancing blows from hurricanes or are too small to appreciably diminish their strength. In slowly moving storms, the rainfall can be of biblical proportions. La Reunion Island

Figure 24.9: Total rainfall from Hurricane Floyd of 1999.

▶

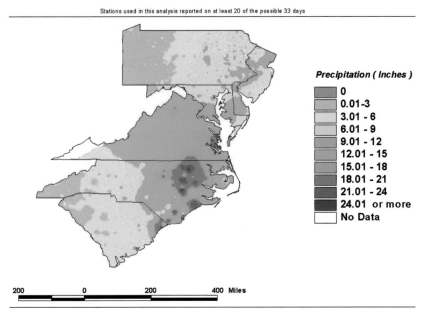

Precipitation Analysis During the Period August 29-September 30th, 1999

Stations used in this analysis reported on at least 20 of the possible 33 days

Precipitation (Inches)

0
0.01-3
3.01 - 6
6.01 - 9
9.01 - 12
12.01 - 15
15.01 - 18
18.01 - 21
21.01 - 24
24.01 or more
No Data

200   0   200   400 Miles

in the South Indian Ocean has the distinction of having broken rainfall records in each of five categories. One storm, in January 1966, produced 1.144 m (45") of rain in 12 hours, and 1.825 m (71.8") of rain in 24 hours. Another, in April 1958, broke the 48-hour rainfall record with 2.467 m (97.1") of rain. A third storm performed a loop near La Reunion in January 1980, yielding an unimaginable total of 5.678 m (223.5," or more than 18.5 ft) of rain over ten days.

Figure 24.10: Franklin, Virginia, in the aftermath of Hurricane Floyd of 1999.

▶

Even where the terrain is flat and dry, tropical cyclones can produce enormous rainfall when they interact with other weather systems. A case in point is Hurricane Floyd of 1999, which produced serious flooding over a large swath of the U.S. piedmont region. In this case, a preexisting frontal zone stretched along the path of Floyd, with cold air to the northwest and warm, very moist air to the southeast. As the rapidly weakening storm moved northeastward along this frontal zone, the strong southeasterly flow ahead of it was forced to ascend over the colder air, as if the cold air were a mountain range. The result was prolonged, extremely heavy rain well after Floyd made landfall in North Carolina, as shown in Figures 24.9 and 24.10. Most of the rain fell over the coastal plain and, in this case, did not appear to be strongly enhanced by the Appalachian Mountains.

Hurricanes still register in our minds as ferocious wind storms, and we readily leave coastal areas when warned of their approach. But some of the highest hurricane death tolls result from inland flooding, which is more difficult to predict and whose danger is still underappreciated. While forecasters emphasize the danger of flooding, they continue to rate storms by their maximum winds, and media attention is focused almost exclusively on wind, waves, and storm surge. Reduction of death and injury from inland flooding will require both better education about its true dangers and improved rainfall forecasts.

▲

Ships in Distress in a Storm. Oil painting by British artist Peter Monamy, ca. 1720–30.

I came into a place mute of all light,
Which bellows as the sea does in a tempest,
If by opposing winds 'tis combated.

The infernal hurricane that never rests
Hurtles the spirits onward in its rapine;
Whirling them round, and smiting, it molests them.

—Dante (1265–1321), from the *Divine Comedy*,
translated by Henry Wadsworth Longfellow

25   The Hunters

In my opinion, a hurricane is not the place in which to fly an airplane.

—Lieutenant Redding W. Bunting, B-25 navigator, September 1945

The idea of flying an airplane into the violence of a hurricane strikes most people as about as rational as going over Niagara Falls in a barrel. But most flights into hurricanes are not very rough. If you are a frequent flyer, chances are you have been in worse turbulence on a commercial airliner than is normally encountered in hurricanes. Of course, there have been exceptions.

It must have taken enormous courage to do it the first time. The technical advance that made this breakthrough possible was instrument flying. Up through the early 1940s, pilots relied strictly on visual cues and a compass to fly and navigate their aircraft. Flying into even the most benign cloud without instruments is very hazardous, since pilots lose those visual cues that enable them to keep flying level and straight. In the early days of flight, pilots would often navigate by following highways and railroads.

Among the pioneers of instrument flying was Joseph B. Duckworth, an Eastern Airlines pilot who resigned from the airline to join war preparations in 1940, signing on as an Army Air Corps major. Recognizing the enormous advantages of being able to fly in any kind of weather, Duckworth helped establish an instrument flying instructors school, in Bryan, Texas, in February 1943. He rapidly gained a reputation for teaching young pilots to fly by instruments in all kinds of weather.

On the morning of July 27, 1943, Duckworth heard that a hurricane was approaching Galveston, about 120 miles southeast of Bryan. In residence at the time was a group of British pilots, many of whom had flown combat missions over Europe. Taunted by the Brits, Duckworth proposed to fly into the storm and recruited for the purpose a reluctant Lieutenant, Ralph O'Hair, as navigator. Telling no one of his plans, he managed to fly an AT-6 trainer into the eye of the storm. He and O'Hair became the first aviators to penetrate the core of a hurricane, the first airborne witnesses of a

Figure 25.1: An AT-6, like the plane used by Duckworth to penetrate the core of a hurricane in 1943.

▲

hurricane's eye. On returning to base, Duckworth was approached by the field's weather officer, who was so disappointed at having been left behind that Duckworth felt obliged to take him back to the storm. When word of these flights got around, several others decided to fly into the storm in twin-engine B-25s.

Duckworth's flight showed that even a small aircraft could penetrate a hurricane. He later recounted that his only real concerns were loss of communications owing to static and the possibility that heavy rain might cause engine failure. At that time, loss of radio also meant loss of navigation, which depended on radio signals.

Soon after this first mission, several more flights were conducted into Atlantic hurricanes. In February 1 944, the Army Air Corps, the Navy, and the Weather Bureau established a plan for airborne hurricane reconnaissance, resulting in several missions during the 1944 hurricane season. Thus was inaugurated the era of routine hurricane reconnaissance, which continues in the Atlantic region to this day.

In the early years of airborne reconnaissance, forecasters would request flights based on limited data, mostly from islands and ships. The aircraft crews would usually set out with only a very approximate estimate of the position of the storm and had to rely on their own visual observations to guide them to the storm center. They estimated the speed and direction of the wind by looking at the sea surface. If clouds obscured the surface, there was little to guide the pilots in their hunt for the storm center.

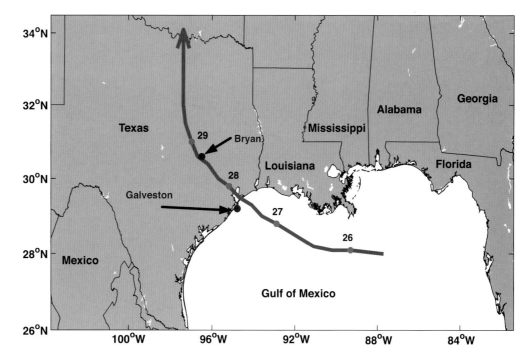

Figure 25.2: Track of the Galveston Hurricane of 1943, the first storm to be penetrated by aircraft. The location of Duckworth's base in Bryan, Texas, is indicated.

▶

Before the advent of meteorological satellites in the 1960s, forecasters relied on reconnaissance aircraft for precisely locating storms at sea. With experience, pilots became more skilled at pinpointing storms and, in consequence, forecasts improved. One ironic result of the improved forecasts was that ships were better able to steer clear of violent storms—yielding fewer ship observations near storm centers.

Not all reconnaissance flights are uneventful. Flights into rapidly developing storms in particular can be extremely rough, particularly at low altitudes. In the early days, the only way of estimating the wind direction and speed was to observe the ocean surface, requiring pilots to fly only a few hundred feet above it. The hurricane airflow is particularly turbulent in the first few thousand feet above the surface, in a region called the atmospheric boundary layer. Here, friction between the rapidly flowing air and the rough underlying surface leads to intense turbulent eddies that can throw an airplane around like a leaf in an autumn windstorm. Lieutenant James P. Dalton describes an early Navy B-25 hurricane mission, flown at very low altitudes into an Atlantic hurricane in September 1945:

> Frankly speaking, throughout my entire life I have been frightened, really frightened, only three times. All of this was connected intimately with weather reconnaissance. I think I can truthfully and without exaggeration say that absolutely the worst time was while I was flying through Kappler's Hurricane on September 15, 1945. We were stationed at Morrison Field, West Palm Beach, Florida, at the time. Everyone except the Duck Flight Recco Squadron had evacuated the field for safer areas the day before.
>
> Hurricane reconnaissance being our business, we of course stayed on, in order to operate as closely as possible to the storm. We were to take off at 7:00 A.M. local time and by then several thunderstorms had already appeared, thoroughly drenching us before we could climb into our plane. But each crew member was keenly alert, for he knew what to expect. I've flown approximately fifteen hundred normal weather reconnaissance hours; that is, if you can call going out and looking for trouble "normal

flying." I have covered the Atlantic completely north of the equator to the Arctic Circle, flying in all kinds of weather and during all seasons but never has anything like this happened to me before.

One minute this plane, seemingly under control, would suddenly wrench itself free, throw itself into a vertical bank and head straight for the steaming white sea below. An instant later it was on the other wing, this time climbing with its nose down at an ungodly speed. To ditch would be disastrous. I stood on my hands as much as I did on my feet. Rain was so heavy it was as if we were flying through the sea like a submarine. Navigation was practically impossible. For not a minute could we say we were moving in any single direction—at one time I recorded twenty-eight degrees drift, two minutes later it was from the opposite direction almost as strong. But then taking a drift reading during the worst of it was out of the question. I was able to record a wind of 125 miles an hour, and I still don't know how it was possible, the air was so terribly rough. At one time, though, our pressure altimeter was indicating twenty-six hundred feet due to the drop in pressure, when we actually were at seven hundred feet. At this time the bottom fell out. I don't know how close we came to the sea but it was far too close to suit my fancy. Right then and there I prayed. I vouched if I could come out alive I would never fly again.

Low-level penetrations into hurricanes and typhoons were routine from the mid-1940s through the late 1950s. During this era, several aircraft were lost in tropical cyclones. The first known loss was of a Navy PBY on October 1, 1945, flying in a typhoon over the South China Sea. On October 26, 1952, a B-29 with ten crew members disappeared without a trace during a low-level penetration of Typhoon Wilma some three hundred miles east of Leyte. The last words from the crew were "Getting rough." A year later, on December 16, 1953, a Navy PBY aircraft crashed while making a low-level flight into Typhoon Doris in the western Pacific, killing all nine crew members.

A P2V Neptune was lost with all hands in the Caribbean, in 1955, while penetrating the eye of Hurricane Janet at a last reported altitude of seven hundred feet. Again, no wreckage was ever found. This was the only time a hurricane hunter aircraft was lost during an Atlantic reconnaissance mission. But there have been close calls. On August 23, 1964, a Navy WV-2, a military version of the then-common Constellation airliner, was flying into Hurricane Cleo, about a hundred miles south-south-west of Puerto Rico. On entering the eyewall, the plane was gripped by extreme updraft turbulence, which ripped away the left wing-tip fuel tank and part of the left wing. On trying to fly out of the eye, the aircraft encountered extreme turbulence in a downdraft, shearing off the right wing-tip fuel tank and an even larger section of the wing itself. Somehow, the pilot managed to get the plane back to an airfield, but many of the crew were severely injured, and the aircraft itself was so badly damaged that it had to be scrapped.

On January 15, 1958, a WB-50 hurricane hunter was lost with all hands during a flight into Typhoon Ophelia, southwest of Guam. Despite an extensive search, no trace of the airplane or its crew was ever found.

The most recent fatal reconnaissance mission occurred in October 1974, when a U.S. Air Force C-130 aircraft and its crew of six were lost in a typhoon over the South China Sea.

One of the hazards of low-level hurricane penetrations in the early days was the absence of reliable techniques for determining aircraft altitude. Pilots had to rely on a pressure altimeter, which, as its name implies, actually measures atmospheric pressure. Normally, pressure varies far more rapidly

Figure 25.3: A U.S. Navy WV-2, just after returning from Hurricane Cleo in 1964. Note the damaged wing tip where a fuel tank once was. Several crew members were severely injured.

▲

in the vertical than it does in the horizontal, so pressure is a good proxy for altitude. But the very low surface pressures near the centers of hurricanes "fool" the altimeter into reading comparatively high altitudes. Aviators were perfectly aware of this problem, but without a reliable altitude measurement, they were forced to rely on visual observation of the sea surface.

Beginning in the late 1950s, several technical advances made it feasible to make useful measurements from higher altitudes. An early advance was the radar altimeter. By sending pulses of electromagnetic radiation down to the surface and measuring the time for backscattered radiation to return to the aircraft, one can obtain a precise measure of altitude. Later, Doppler radars were pointed down and forward, at an angle to the surface. These new radars not only measure the intensity and time of return of the backscattered radiation, they also measure the Doppler shift, which shows how fast the aircraft is moving over the ground. The speed of the aircraft through the air is also measured, and the vector difference between these two measures yields the speed and direction of the wind relative to the ground. With Doppler radar, pilots no longer had to see the ocean to estimate the wind speed.

In 1949, Charles Stark Draper demonstrated the first inertial navigation system. Draper, a professor at the Massachusetts Institute of Technology, recognized that precisely manufactured gyroscopes could measure accelerations with enough accuracy to determine the velocity and position of ships and aircraft. By the late 1960s, inertial navigation systems based on these gyroscopes were being

Figure 25.4: Variation with distance from storm center of (top) the horizontal wind velocity and (bottom) the vertical air velocity in a computer model storm that has just undergone rapid intensification. Note how quickly the wind speed rises, from only about 5 mph at 12 mi radius to 85 mph at 25 mi. Between 20 and 25 mi, there is also a big increase in updraft speed. A NOAA-P3 may have flown through a wind field like this in Hurricane Hugo of 1989.

▶

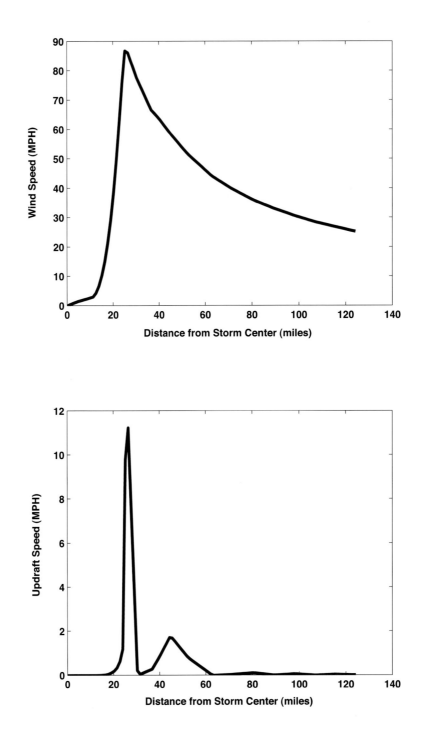

installed in military aircraft. Inertial navigation was a great boon to hurricane reconnaissance, as it not only allowed the navigator to determine the aircraft's position with great accuracy, but also gave its speed and direction of movement over the surface. This, together with conventional measures of airspeed, gives the speed and direction of the wind in which the aircraft is flying. Beginning in the 1990s, inertial navigation was supplemented by the Global Positioning System, which pinpoints location using signals from a constellation of satellites.

As instrumentation improved, it became increasingly common to penetrate hurricanes at altitudes well above the atmospheric boundary layer, where the air is less turbulent, even though the actual winds speeds are often greater. (By the same token, a Los Angeles–to–New York flight can benefit from a 150 mph tailwind while its passengers enjoy a peaceful ride.) What jolts an airplane is not the wind per se, but variations in wind across the airplane. For this reason, thunderstorms, which have strong updrafts and downdrafts only a few thousand feet apart, are very hazardous to aircraft, and pilots take great care to avoid them.

The real hazard of hurricane reconnaissance is the storm's eyewall, where strong updrafts can give way to equally strong downdrafts in a few hundred feet. Worse yet, the horizontal wind speed can change by more than 100 mph in a few seconds of flight time as the aircraft flies across the inner edge of the eyewall, emerging into the eye. This can put severe strains on the airplane and its crew.

In recent years, we have learned through theory, computer modeling, and hard experience that the inner edge of the eyewall of a storm that has just intensified rapidly can be especially hazardous. When a hurricane intensifies, the eyewall wind velocities accelerate while the eye itself tends to remain calm. The eyewall becomes an atmospheric front, across which there are very sharp changes of wind (Figure 25.4). (An eddy with a peak wind speed of 104 m/s [233 mph] was measured in Hurricane Isabel in 2003.) The enormous wind shear across the eyewall is highly unstable and breaks down into small but extremely intense eddies, some of them tornado-like in their size and ferocity. These eddies mix momentum into the eye, smoothing out the front-like jump in velocity and rendering the eyewall less hazardous to aircraft. It takes many hours to accomplish this, however, and if the storm is intensifying rapidly, the wind shear will build up more rapidly than it can be dissipated.

Figure 25.5: A U.S. Air Force C-130 hurricane hunter.

▶

Figure 25.6:

A NOAA WP-3D

reconnaissance aircraft.

▶

A near tragedy occurred in September 1989, when a NOAA P-3 Orion penetrated the eyewall of Hurricane Hugo over the central North Atlantic, just after the storm had intensified rapidly. In a few seconds of flying, the speed of the wind through which the aircraft was flying dropped from over 175 mph to less than 20 mph across the inner edge of the eyewall. Enormous accelerations wreaked havoc in the cabin, and a fuel valve failed on one of the four engines, sending flames roaring out of the exhaust and forcing the pilots to cut power to the engine. With nearly full tanks and only a few thousand feet of altitude, the pilot made a calculated gamble and dumped fuel, in spite of the engine fire. The gamble paid off, and the plane slowly climbed while circling in the eye, later making a safe return to base.

In spite of the sometimes harrowing events like these, the vast majority of hurricane reconnaissance missions are uneventful, and the knowledge gained has been of incalculable value. Data collected during these missions have led to a detailed picture of the structure and behavior of hurricanes (see Chapter 2) and some understanding of how they develop, intensify, and move.

Today, hurricane reconnaissance aircraft carry sophisticated equipment for measuring properties of the atmosphere and the upper ocean. In addition to measuring standard meteorological quantities such as wind, temperature, humidity, and pressure, many reconnaissance flights deploy dropwindsondes, instrument packages on parachutes that float down to the sea surface, transmitting information back to the parent aircraft. Atmospheric sensors on the sonde measure temperature, humidity, and pressure, while an onboard GPS chip keeps track of the sonde's exact location and drift, from which the wind direction and speed can be deduced. Data collected from the aircraft and the sondes are immediately transmitted back to the mainland United States and used in computer models of hurricane motion, substantially improving forecasts of the storm's movement.

Figure 25.7: The newest member of the NOAA weather fleet: a Gulfstream-IV jet.

►

Reconnaissance aircraft also occasionally deploy Airborne Expendable Bathythermographs (AXBTs), small floats or buoys that are dropped to the sea surface on parachutes. Once afloat, they deploy long wires with thermistors that measure ocean temperature down to depths of a hundred meters or so. These measurements are transmitted to the aircraft by radio.

The collection of data for research and forecasting is today the primary task of the hurricane reconnaissance mission, the role of "hunter" having been long ago obviated by meteorological satellites that accurately locate tropical disturbances all around the globe.

Although the safety record of hurricane reconnaissance has steadily improved, the U.S. military discontinued routine reconnaissance missions over the Pacific in 1987, both for fiscal reasons and because it was felt that satellites could provide enough data to forecasters. There is some evidence that the lack of direct measurements in storms outside the Atlantic significantly degrades estimates of the storms' strength and structure, thereby yielding inferior forecasts.

A new tool that is gradually becoming available for hurricane reconnaissance is the robotic airplane, also known as an unmanned aerial vehicle (UAV) or remotely piloted vehicle (RPV). These have been used in military reconnaissance for many decades, and more recently have engaged in armed combat. Remotely piloted vehicles have been developed for environmental monitoring. For example, the aerosonde, a lightweight RPV about the size of a large model aircraft, can be easily deployed from the top of a car and has enough endurance to cross the Atlantic. Such aircraft could potentially monitor hurricanes, including their boundary layers, which are often considered too hazardous for piloted aircraft. At the other end of the spectrum are the larger military RPVs, some of which are being modified for weather reconnaissance. With their unusually high maximum operating

Figure 25.8: A U.S. Air Force Global Hawk remotely piloted vehicle on a test flight. Robotic aircraft such as these may revolutionize weather reconnaissance.

▶

altitudes, such aircraft could fly over hurricanes, deploying lightweight dropwindsondes to measure meteorological quantities within the storm, all the way from the top of the circulation right down to the sea surface. These new tools may lead to a resurgence of airborne hurricane reconnaissance around the world.

▲

Hurricane Frederick,
Mobile, Alabama. Casein
painting by American artist
Howard Worner, 1980.

The harbor lights have long gone out,
the town is sound asleep.
Like omens underneath a moon,
the waves crash on the beach.
The restless clouds are circling
like birds of prey in flight.
And Nature lifts an angry hand,
poised and set to strike.

And every man will know the power
that marked the planet's birth.
The hurricane, the hurricane
returns us to the earth....

No man-made shelter strong enough
to stop the raging tide.
The storm relentless in its quest
to conquer and divide.
What has stood a hundred years
awakens to the roar.

As the waves come pounding down
like hammers on the shore....

And every man will know the power
that marked the planet's birth.
The hurricane, the hurricane
returns us to the earth....

The harbor lights have long gone out,
submerged beneath the waves.
The moon attends the final rites
above the ocean grave.
While restless clouds still circle 'round
like birds of prey in flight,
Nature's hand, all quiet and still,
retreats before the light....

The hurricane is over now,
the storm has finally passed.
While on the sand a child walks
and kicks at bits of glass....

> —"The Hurricane (Hammers on the Shore),"
> a song by Peter Schilling (b. 1956)

26 Hurricane Camille

*Many were warned to evacuate. Some refused. Some did leave and
returned. Everyone thought their houses or buildings were safe because
they'd survived the last bad hurricane 22 years earlier.*

—Julia Guice, Civil Defense director for Biloxi, Mississippi

On the night of Sunday, August 17, 1969, Hurricane Camille, one of the most violent storms
ever to strike the United States, obliterated the Gulf coast of Mississippi, taking at least 150 lives and
destroying whole towns and fishing fleets. Though the winds died down quickly after landfall, the circulation and its excessively wet core drifted northward and then eastward, where it encountered the
Appalachian Mountains, precipitating a series of deluges that killed another 150 people. A close look
at the meteorological circumstances suggests that, like many great disasters, Camille was the result of
the confluence of several improbable events.

On August 5, forecasters noticed an easterly wave moving off the coast of Africa. Over the next
nine days, the wave tracked westward across the tropical Atlantic, with little change, but on the fourteenth, a Navy reconnaissance pilot recorded a central pressure of 999 mb (29.50") and surface
winds of 25 m/s (55 mph). The next day the storm, now named Camille, reached hurricane strength,
and late in the afternoon, as it approached Cuba, it was packing 50 m/s (115 mph) winds. Forecasters warned that Camille was the most intense Atlantic hurricane since Beulah of 1967. During the
evening of the fifteenth, it crossed the western tip of Cuba, bringing 40 m/s (90 mph) winds and 25
cm (10 in) of rain.

During August 16, Camille turned north by northwest and intensified rapidly, heading into the
eastern Gulf of Mexico. Early that day, the National Hurricane Center issued a hurricane watch from
Biloxi, Mississippi, to St. Marks, Florida. Until about this time, the storm had been under almost continuous surveillance by reconnaissance aircraft, which forecasters relied on for intensity estimates. But
on the sixteenth, both available aircraft were out of service for maintenance. Dr. Robert Simpson, then
the Center's director, persuaded the U.S. Navy to attempt a reconnaissance mission using their only

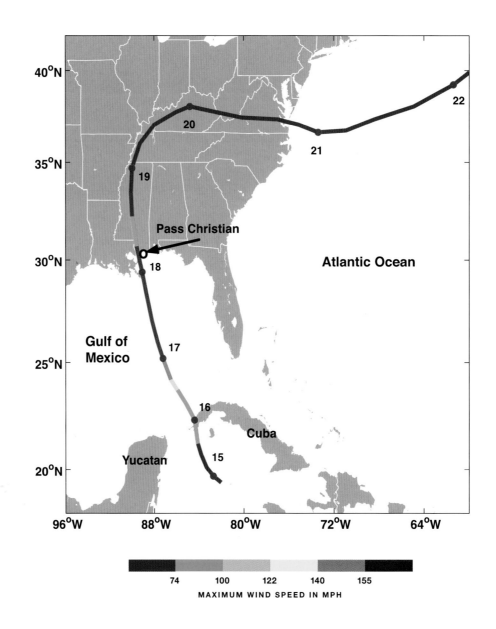

available aircraft—one not particularly well suited to working in hurricanes. The crew flew to within a hundred miles of Camille's center, but then turned back, not wanting to risk damage to the aircraft. Although the aircraft's radar showed that Camille had a well-defined eye, little could be deduced about the storm's true intensity. Increasingly worried that Camille was becoming a dangerous hurricane, Simpson begged the Air Force to redirect some of its aircraft to fly Camille. The Air Force complied, and by later in the day, a reconnaissance aircraft was able to report an accurate storm position and, more ominously, winds that had increased to 65 m/s (150 mph).

Camille moved relentlessly onward toward Louisiana and Mississippi. On the morning of Sunday, August 17, hurricane warnings were extended from the Florida panhandle westward to New Orleans. Early that afternoon, a reconnaissance aircraft flew into the storm center and reported back a phenomenally low central pressure of 901 mb (26.61"), with maximum winds around 90 m/s (200 mph). Camille was now the most intense hurricane on record since the Labor Day storm of 1935.

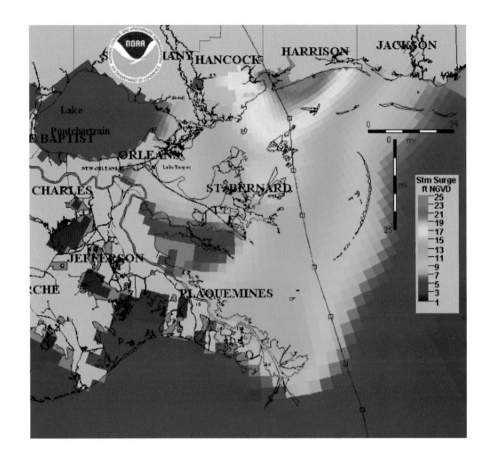

Figure 26.2: Storm surge during Hurricane Camille, estimated using a computer model. The model's predictions agree well with tide gauge measurements.

▶

Radio and television warnings during that Sunday were dire, and though most people in the path of the storm evacuated, some obstinately remained. As a rule, the media must decide how much emphasis to place on warnings, and in this case, while most of the local television and radio stations stressed the critical need to evacuate immediately, others were less emphatic, interspersing the Weather Service warnings with regular programs and advertisements. Under U.S. state and local laws, it is sometimes difficult to mandate an evacuation, and only in a few cases did police go to the extreme of arresting and forcibly evacuating those who had decided to remain in the path of the storm. To make matters worse, the fast approach of Camille and heavy winds and rains ahead of its core flooded many of the bridges spanning the creeks and rivers of the region's low-lying terrain, making escape difficult or impossible.

Shortly before midnight on Sunday, August 17, Camille stormed ashore near Bay St. Louis, Mississippi, with an estimated central pressure of 909 mb (26.84") and accompanied by winds of 85 m/s (190 mph). The intense winds and the 15 mph forward movement of the storm created an immense storm surge, measuring 22.6 ft at Pass Christian, 21.6 ft at Long Beach, 21 ft at Gulfport, 19.5 ft at Biloxi, and 15 ft on Biloxi Bay. It was the highest storm surge ever recorded in the United States (Figure 26.2).

The strongest winds struck east of the eye, between Pass Christian and Long Beach, Mississippi, where destruction was almost complete (Figure 26.3). Among those who had risked riding out

Figure 26.3: Downtown Pass
Christian, Mississippi, after
Camille. City Hall is in the
background on the left.

▶

the storm were 24 residents of a beachfront apartment complex at Pass Christian, whose "hurricane party" was interrupted by rapidly rising water and the rapid disintegration of the building. Only three of the revelers survived; one of those was found the next day clinging to the top of a tree, five miles inland. Ships of all sizes were carried inshore and smashed among the wreckage of buildings and vegetation. As Camille moved inland, it left at least 150 dead in its wake.

Camille was something of a meteorological freak, as revealed by a computer simulation of the storm's well-observed intensity (Figure 26.4). The simulation failed dramatically. Although winds reached a respectable 60 m/s (140 mph), they did not approach the observed maximum winds that exceeded 85 m/s (190 mph). This failure was likely due to the presence along the storm path of a narrow and meandering current of warm ocean water known as the Loop Current.

During hurricane season, the surface waters of the Gulf of Mexico normally have temperatures around 85°F, with little variation across the Gulf. In most places, the warm water occupies only a shallow layer, less than a hundred feet deep. Underneath this warm layer lies much colder water. Most hurricanes passing over the Gulf churn up this colder water, reducing the surface temperature under the critical eyewall region, and thereby preventing the storm from becoming very strong (see Chapter 12). Our computer simulation assumed that hurricanes move over this climatologically normal Gulf.

In reality, the thickness of this warm water layer varies quite a bit and is particularly thick in the Loop Current, which enters the Gulf through the Yucatan Straits, penetrates some distance northward into the Gulf, and then loops eastward and southward, parallel to the west coast of Florida, finally exiting eastward through the Straits of Florida, where it is renamed the Gulf Stream. This current is unstable, and meanders back and forth, sometimes extending northward almost to the Mississippi delta,

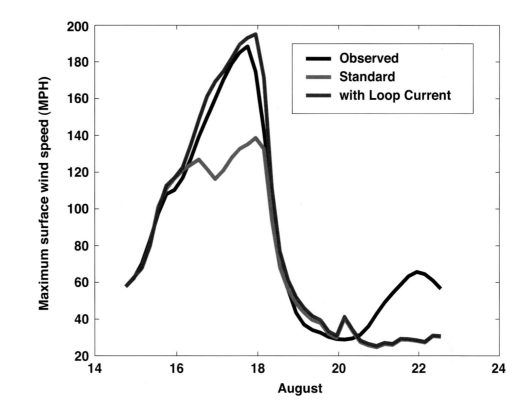

Figure 26.4: Observed evolution of Camille's maximum wind speeds (blue curve), together with two computer simulations: a simulation using the standard Gulf of Mexico upper-ocean conditions (green curve) and one using upper ocean temperature typical of the Loop Current (red curve).

▶

sometimes barely penetrating the Gulf. It changes from month to month, often doubling back on itself and forming warm, clockwise-spinning eddies that drift slowly westward, crossing the Gulf over a period of several months.

Unfortunately, there are no routine measurements of this current. Measuring the subsurface ocean is difficult and expensive, requiring ships and aircraft from which measurements of properties such as temperature and salinity are made. To make matters worse, the Loop Current is usually invisible in the surface temperature, which is quite uniform in summer. We have no good measurements to tell us where the Loop Current was at the time of Camille. But detailed surveys of the Gulf have been done at other times. Figure 26.5 shows an east-west cross-section of temperature across the central Gulf, made in August 1965. The Loop Current can be seen at the far right of this figure, where warm water penetrates to much greater depth than elsewhere.

The Loop Current is normally quite narrow—usually only about 80 km (50 mi) across—and a hurricane would normally take only a few hours to cross it. Only hurricanes moving nearly along the current's axis would be significantly affected by it. We assumed that this was the case in Hurricane Camille and used the temperature structure at the far right of Figure 26.5 in a second computer simulation of Camille, as shown in Figure 26.4. The new simulation is much better. In fact, the only way we can simulate Camille's extraordinary power is to assume that it moved right up the axis of the Loop Current, a very improbable occurrence given the current's narrow width. Had Camille moved along a track 50 mi either side of its actual path, it would probably have been markedly less powerful.

Camille weakened rapidly as it moved northward through Mississippi, with central pressures rising to 980 mb (28.90") and wind gusts to just 30 m/s (67 mph) early on Monday the eighteenth. Winds stronger than 45 m/s (100 mph) had been experienced only within about 160 km (100 mi) of the coast. Rainfall, which exceeded 28 cm (11 in) in parts of southern Mississippi, also diminished rapidly, with storm totals of 5–11 cm (2–4.5 in) through most of western Tennessee, Kentucky, and West Virginia.

But then, without warning, Camille came back to life as it moved eastward into Virginia. Around 7 P.M. on Tuesday, August 19, torrential rains lasting more than eight hours caused flash floods and landslides along the eastern slopes of the Blue Ridge Mountains and record flooding in Virginia's James River watershed. Around midnight, Camille intensified as it passed south of Roanoke and Lynchburg. Rainfall increased rapidly to the northeast of the low-pressure center along the western slopes of the Blue Ridge Mountains, with more than 10 inches recorded at Clifton Forge, Virginia. In Nelson County, Virginia, Camille dumped 31 inches of rain in six hours. Residents remembered the rain as resembling a massive waterfall. The rainfall continued to increase on the eastern slopes of the Blue Ridge Mountains, reaching catastrophic proportions.

Figure 26.5: West-east cross section through the central Gulf of Mexico made using instruments deployed from ships, in August 1965. Contour lines are drawn at intervals of 1°C, with the coldest water at 10°C and the warmest water at 29°C. The Loop Current is evident at the far right side of the diagram. After Leipper (1967).

▶

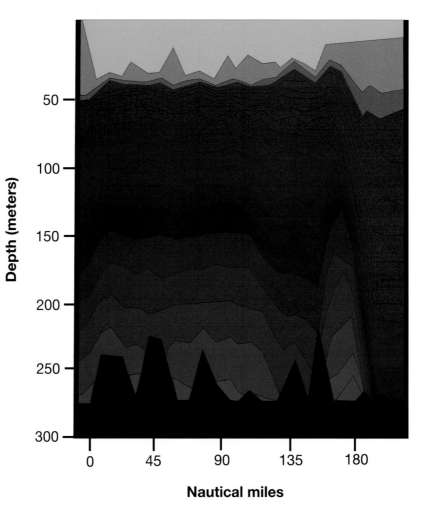

These flash floods and landslides constituted the worst natural disaster ever to affect the state of Virginia. Most residents were asleep during the storm, which had been unpredicted. Even if forecasters had realized what was happening, there was no way to warn anyone of the impending catastrophe, because phone lines were obliterated along with everything else by flash floods barreling down creeks and rivers.

At least 153 Virginians lost their lives in Camille; of these, 126 were residents of Nelson County. This represented a little more than 1 percent of that county's population. After the storm, only one highway in Virginia remained intact.

On Wednesday afternoon, August 20, Camille moved off the coast east of Norfolk and regained tropical storm intensity. Two days later, it merged with a frontal system and lost its identity as a tropical storm some 280 km (175 mi) southeast of Cape Race, Newfoundland. It had killed at least three hundred people and done billions of dollars in damages, leaving in grief a nation that just one month earlier had put the first man on the moon.

▲

The Slave Ship (Slavers Throwing Overboard the Dead and the Dying, Typhoon Coming On). Oil painting by James Mallord William Turner, exh. 1805.

Aloft all hands, strike the top-masts and belay;
Yon angry setting sun and fierce-edged clouds
Declare the Typhoon's coming.
Before it sweeps your decks, throw overboard
The dead and dying—; ne'er heed their chains
Hope, Hope, fallacious Hope!
Where is thy market now?

　　—James Thomson (1700–1748),
　　　"The Seasons" (Turner attached these lines
　　　　to his painting at its original exhibition.)

27   Into the Maelstrom: A Photo Essay

The eye was surrounded by a coliseum of clouds whose walls...were banked like galleries in a great opera house.

—Robert H. Simpson

What is it like to fly into a hurricane? No words or pictures can fully convey the terrible three-dimensional beauty of this natural wonder; one simply has to go there. On these pages are some of the more striking photographs taken on reconnaissance missions together with the words of one of those privileged few witnesses of this great natural spectacle. These photos were taken in many different storms, but have been arranged to convey a typical sequence of scenes one might encounter on a hurricane reconnaissance mission.

The first sign of the storm are the outer bands of cumulus and cumulonimbus, aligned with the surface wind, and high, icy cirrus flowing out of the storm's core.

A last look at the outside world: As one flies in toward the storm center, the thin cirrus are gradually replaced by a high, dense overcast of altostratus. This view, looking back out, away from the storm center, catches some of the outer bands.

Rain showers become more and more frequent, and the sea shows the first signs of high wind, with breaking waves and white patches of bubbles. Here the wind speed is about 80 mph.

As one closes in on the eyewall, the view often looks like this, with thick clouds reducing visibility to a few feet and heavy rain streaking across the window.

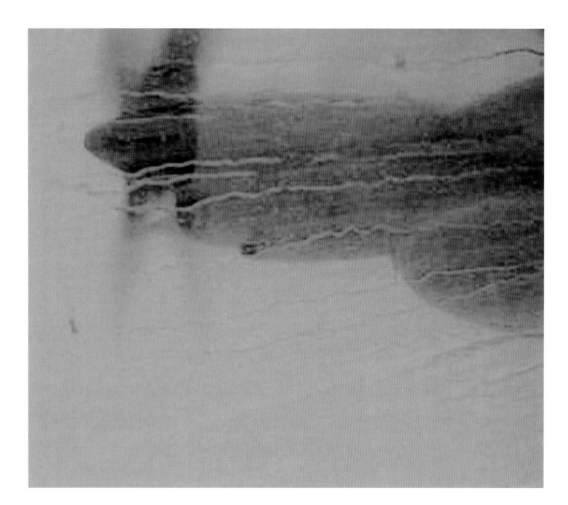

No picture can do justice to the sensation of being aloft in the eye of a hurricane. This photo of the eye of Hurricane Georges of 1998 shows a ten-mile-high stadium of white eyewall cloud partially illuminated by the setting sun. As is often the case, the base of the eye is mostly filled with shallow cumulus and stratocumulus, though the sea surface can often be discerned through breaks in the undercast. The tops of these clouds are usually a few thousand feet above the ocean surface. The thin wisps of cloud just above the rim of the eyewall, at upper right, are small ice crystals cascading down the inside of the eyewall, like a waterfall of stardust.

Dr. Robert H. Simpson, who was one of the first scientists ever to fly into a tropical cyclone and who later became the director of the National Hurricane Center in Miami, wrote this vivid description of the eye of a hurricane:

> As we came closer, the surface winds grew stronger. Two hundred miles from the center they reached hurricane force—74 miles per hour—and in another 50 miles they had increased to 100 miles per hour. From here on we could no longer see the surface, for the cloud cover now engulfed the plane completely. Only the spiral pattern of the squall lines on the radar screen enabled us to keep headed toward the storm center. Soon the edge of the rainless eye became visible on the screen. The plane flew through bursts of torrential rain and several turbulent bumps. Then suddenly we were in dazzling sunlight and bright blue sky.
>
> Around us was an awesome display. Marge's eye was a clear space 40 miles in diameter surrounded by a coliseum of clouds whose walls on one side rose vertically and on the other were banked like galleries in a great opera house. The upper rim, about 35,000 feet high, was rounded off smoothly against a background of blue sky. Below us was a floor of smooth clouds rising to a dome 8,000 feet above sea level in the center. There were breaks in it which gave us glimpses of the surface of the ocean. In the vortex around the eye the sea was a scene of unimaginably violent, churning water.

An idea of the state of the sea under the eyewall of an intense hurricane may be gleaned from this remarkable photograph of the inner edge of the eyewall of Hurricane Gilbert of 1988, which set an all-time record for the lowest pressure observed in an Atlantic hurricane.

The center of the vortex is off the lower left corner of this picture. Traveling outward from the center (starting in the lower left and proceeding upward and rightward in this picture), there is a sudden increase in wind speed, revealed by a sheet of sea spray blasted upward from the ocean. The pancake-shaped white patches are places where the wind is literally scooping water from the sea; these are large enough to contain a ship. The eyewall itself can be seen in the upper part of the picture, nearly touching the sea surface. This is a place of unmitigated violence, where bubble-filled water gradually gives way to spray- and cloud-filled air, with no definite interface that one could call the ocean surface. It is arguably the worst place in the world for a ship.

This view of a typhoon eyewall evokes a scene from Dante's Inferno:

I came into a place mute of all light,

Which bellows as the sea does in a tempest,

If by opposing winds 'tis combated.

The infernal hurricane that never rests

Hurtles the spirits onward in its rapine;

Whirling them round, and smiting, it molests them.[18]

<hr />

[18] From The *Divine Comedy*

This remarkable photo looks straight up the eye of a typhoon and was taken from an Air Force C-130 aircraft in a steep turn.

The setting sun guides a homeward-bound hurricane hunter through clear caverns between the outer bands of a storm.

▲

Hurricane Number 27, El Niño, Acrylic painting by French artist Peter Valentiner, 1999.

L ike Rain it sounded till it curved
And then I knew 'twas Wind—
It walked as wet as any Wave
But swept as dry as sand—
When it had pushed itself away
To some remotest Plain
A coming as of Hosts was heard
It filled the Wells, it pleased the Pools
It warbled in the Road—
It pulled the spigot from the Hills
And let the Floods abroad—
It loosened acres, lifted seas
The sites of Centres stirred
Then like Elijah rode away
Upon a Wheel of Cloud.

   —Emily Dickinson (1830–1886)

28

The Great East Pakistan Cyclone of November 1970

The Lord hath his way in the whirlwind and the
clouds are the dust of his feet.

—Nahum 1:3

The Great Hurricane of 1780 is widely known as the most deadly in the Western Hemisphere, having killed more than 20,000 people in the Caribbean. But in a single, horrible night halfway around the world in East Pakistan, a hurricane (known there as a tropical cyclone) killed between 300,000 and 500,000 people—more than ten times the death toll of the 1780 storm—making it the worst tropical cyclone disaster in history.

The region around the mouth of the Ganges River is among the most calamity prone in the world. Consisting of low-lying marshy land and innumerable islands, the coastal plain is highly susceptible to flooding. Inhabitants trying to eke out a living by fishing and farming frequently confront freshwater floods from the Ganges and saltwater storm surges from the tropical cyclones that roar in from the Bay of Bengal, mostly in spring and fall. Owing to the very gentle average slope of the terrain, large areas of land are flooded by relatively small increases in sea level. To make matters worse, the shape of the Bay of Bengal and the gentle shoaling of the waters offshore are conducive to exceptionally large storm surges.

The Bay of Bengal experiences comparatively large normal tide swings, ranging between 3 and 5.5 m (10 and 18 ft). Storms making landfall around the time of low tide generally have little effect, but storms striking near the time of high tide—especially during periods of astronomically high tides—produce devastating floods. When the waters arrive, there is no place to run. With little by way of modern transportation, there is no way to get far enough soon enough to outrun the rising water.

The history of the Ganges delta is thus littered with tragedy. In October 1737, a strong cyclone came ashore near the mouth of the Hooghly River, near Calcutta, plowing a 10–13 m (30–40 ft) storm surge ahead of it and inundating the heavily populated jute plantations of the region. Between

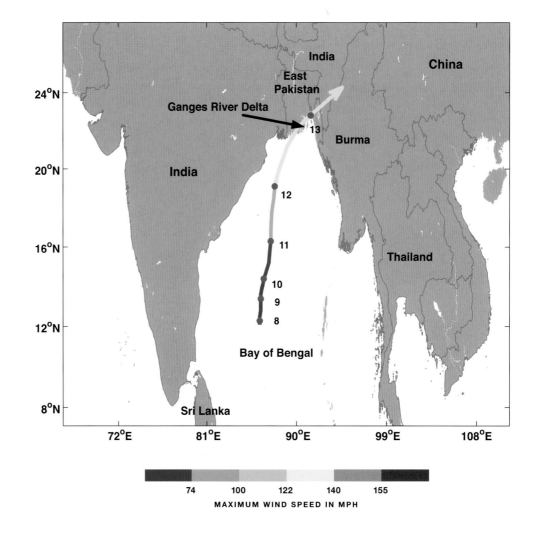

300,000 and 350,000 perished. Tropical cyclones killing more than 10,000 struck again in 1787, 1789, 1822, 1833, 1839, and 1864.

In late October 1876, a particularly vicious tropical cyclone tore through the Ganges delta, accompanied by a 13 m (40 ft) storm surge. The city of Bakergunj was almost completely destroyed, along with nearly all of its livestock. At least 100,000 lost their lives when the storm made landfall, and another 100,000 succumbed to a cholera epidemic in its wake. This was the worst disaster since the Hooghly River storm of 1737. But tragedy struck again in 1897, when another storm took the lives of 175,000. After a merciful lull of 45 years, a Bay of Bengal cyclone killed about 40,000 in October 1942. In this case, World War II concerns interfered with warnings and evacuations, contributing to the high death toll. Another cyclone killed upwards of 22,000 in 1963, and in 1965 three separate storms collectively took the lives of almost 60,000 people.

All this tragedy proved a prelude to the deadliest cyclone in history.

The fall of 1970 was a season of political turmoil in East Pakistan. An election that had been scheduled for October was postponed until December 7, owing to the disruptions caused by heavy

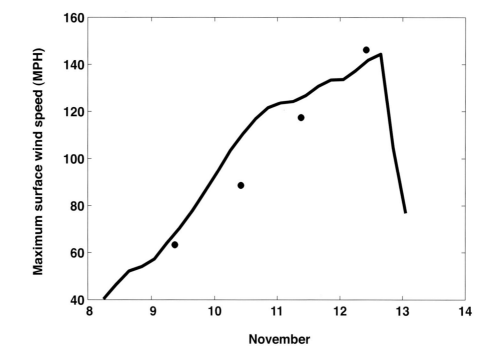

Figure 28.2: Computer simulation of the maximum wind speeds in the Great Cyclone of 1970, based on the observed storm track and estimates of the peak wind on November 8 and 9 from satellite pictures. Dots represent independent estimates based on satellite imagery, made by Dr. Mark Lander.

▶

flooding the previous summer. Pakistan was then a single nation, consisting of West Pakistan (today simply referred to as Pakistan) and East Pakistan, but there was a growing separatist movement in the east, driven mostly by younger people with socialist leanings, belonging to the Awami League, the National Awami Party, and other organizations. It was in this setting that calamity struck.

The Great Cyclone of 1970 had its origins in a tropical storm that had moved westward across Malaysia in early November. Reaching the very warm waters of the Bay of Bengal, the depression began to intensify on the eighth, probably reaching tropical storm strength on the ninth.

Although there were no direct measurements of the winds or pressure in the storm, several photos of the system taken from polar orbiting satellites suggest that the storm intensified rapidly as it accelerated northward toward East Pakistan, arriving at the mouth of the Ganges in the early morning hours of November 13, almost exactly at high tide.

Based on the track of the storm and its probable intensity on the eighth and ninth, as judged from satellite pictures, we have attempted to simulate its intensity. The result (Figure 28.2) suggests average winds in excess of 60 m/s (140 mph) at landfall, consistent with the 60 m/s (140 mph) gusts reported at Chittagong Port, 80 km (50 mi) southeast of the point where the storm center made landfall.

The storm surge of around 4 m (12 ft) occurred almost exactly at the time of high tide, adding another 3 m (10 ft) to mean sea level, to create a total storm surge in excess of 6 m (20 ft).

The storm's toll on East Pakistan's population of 73 million was mind-boggling. Although meteorologists knew of the approaching storm, there was no way to communicate with most of those living in the coastal plain and on the numerous islands in the Ganges delta. On that horrifying night, between 300,000 and 500,000 people lost their lives; many had been asleep when the storm surge struck. Sweeping buildings, vegetation, and dead animals before it, the surge flooded almost a quarter of East Pakistan's landmass. The few who survived this onslaught did so by climbing trees, where they had to battle poisonous snakes. Nearly 46,000 of the estimated 77,000 fishermen of the region perished, and most of the rest were severely injured. About 65 percent of East Pakistan's fishing industry was wiped out, a particularly worrisome outcome since fish were by far the largest source of protein for the country's malnourished residents.

As flood waters receded over the succeeding days, the magnitude of the disaster slowly became evident. Aid began to flow in from many nations, including India and China, but the central government, residing in West Pakistan, did next to nothing. Resentment seethed, and four months later, in March 1971, East Pakistan declared its independence from the west. A bloody ten-month civil war ensued, after which East Pakistan was granted its independence. Thus was born Bangladesh.

The extent of the calamity and its widespread coverage in the international press, replete with gruesome photographs of bloated corpses and grieving survivors, finally prompted governments and private relief organizations to take actions to mitigate the effects of future storms. Given the near impossibility of evacuating large populations to higher terrain, the government and various relief organizations began building cyclone shelters, such as the one shown in Figure 28.3.

Although these were very effective, only 238 were built between 1972 and 1985, financed mostly by the World Bank. After 1985, when a storm killed more than 10,000, another 86 shelters were built. Given that, on average, each shelter can protect roughly 1,000 people, the total sheltering capacity of these structures was around 325,000 by 1991.

Then, in late April 1991, yet another powerful tropical cyclone roared into the region, destroying the city of Chittagong, with 60 m/s (140 mph) winds and a 6 m (20 ft) storm surge, killing 138,000. Many survivors later perished through disease and famine. By that time, the warning and evacuation system had advanced to the point where several million could be evacuated ahead of the storm; otherwise, the death toll would have been even higher.

Over the succeeding six years, various non-government organizations built an additional 1,670 cyclone shelters in Bangladesh, and when a powerful storm hit in May 1997, nearly a million people took refuge in the shelters, and the death toll was limited to about a hundred people.

The advent of satellite detection and computer modeling of tropical cyclones, coupled with modern communication systems, coordinated evacuation plans, and the construction of cyclone shelters, offer the hope that the staggering historical death tolls from Ganges delta cyclones may never be repeated. But history teaches us that a cyclone-free decade or two is all that it takes for a society to forget its past and let down its guard. Where hurricanes are concerned, there is no substitute for constant vigilance.

*Figure 28.3: Contemporary
concrete cyclone shelter in
Bangladesh. Thatched huts
in foreground are typical
of housing in the region; they
offer little protection from
strong winds or floods.*

▲

▲

The Shipwreck. Oil
painting by British artist
James Mallord William
Turner, 1840.

L ook when the clouds are blowing
And all the winds are free:
In fury of their going
They fall upon the sea.
But though the blast is frantic,
And though the tempest raves,
The deep immense Atlantic
Is still beneath the waves.

—Frederick William Henry Myers (1843–1901),
"Wind, Moon and Tides"

29   Forecasting Hurricanes

A woman rang to say she heard there was a hurricane on the way. Well don't worry, there isn't.

—Michael Fish, BBC TV, October 15, 1987. Weather forecast on the night before one of the most violent windstorms ever to strike southern England

A forecaster's heart knoweth its own bitterness
But a stranger meddleth not with its joy.

—Napier Shaw (after Proverbs 14:10)

No natural phenomenon poses a greater challenge to forecasters than the hurricane. Earthquakes cause comparable damage, but at present there is little scientific basis for their prediction. Lightning and flash floods kill more people in the United States than hurricanes, but they evolve so rapidly that forecasters can do little more than predict the conditions that favor them. Major river floods are often devastating, but evolve slowly enough that there is usually ample time for evacuation, and little doubt about where they will occur.

The problem with hurricanes is that we *can* forecast them with some skill, yet they move and evolve fast enough that decisions must be made quickly but with incomplete knowledge. Partial knowledge poses a greater problem than either complete ignorance or full knowledge. If we knew for sure a hurricane was coming, we would get out of the way, however inconvenient. If, on the other hand, hurricanes were completely unpredictable (as they were regarded for millennia), we would take prudent precautions, hope for the best, and get on with our lives. But partial knowledge engenders existential angst: we must carefully consider alternatives, and the decisions can be tough.

Hurricane forecasters are haunted by a handful of nightmare scenarios. For example:

1. ► A weak hurricane is about to cross Cuba into the Straits of Florida, but it has the potential of intensifying dramatically and may hit the Florida Keys in less than two days. With only a two-lane highway leading to safety, it would take at least 36 hours to evacuate the Keys' 80,000 residents. Evacuation will no doubt lead to numerous injuries and possibly death, and will cost many millions of dollars, but if a 150 mph storm hits an unprepared population, the death toll will be staggering.

2. ► A small, rapidly moving hurricane intensifies as it moves northward in the Gulf of Mexico. Its present course will take it ashore along the

Alabama-Mississippi coastline, but a last-minute veer to the left could bring it directly into downtown New Orleans. Much of that city is below sea level and relies on pumps to keep the water out. A strong storm surge coupled with heavy rain could overwhelm the pumps and submerge most residences and one-story buildings. Evacuate, or trust the forecast?

3 ▶ Just north of the Bahamas, a Category 5 hurricane accelerates northward toward Long Island, mimicking the New England Hurricane of 1938. The bulk of the computer guidance suggests that the storm will make landfall in central Long Island, but there is a hint it could curve northwestward around a weak upper-atmospheric trough digging southward over Virginia. Moreover, the guidance indicates landfall near the time of high spring tides, and the Hudson is already high from a summer of unusually heavy rain. To make matters worse, the scheduled balloon sounding of the atmosphere from Buffalo, New York, failed or was not transmitted and in any case was not included in the numerical forecasts, adding to the tense uncertainty. A storm surge model indicates that a 120 mph storm moving rapidly into northeastern New Jersey at the time of high tide could produce a 25 ft storm surge in Manhattan. Evacuate New York City?

Unfortunately, these scenarios are not altogether implausible, given what we know about past hurricane behavior and what research shows to be possible. The decisions forecasters make in these circumstances and far less dramatic ones mean the difference between life and death. Underwarn and thousands of lives are placed at risk. Overwarn and lives are lost in automobile accidents, many millions of dollars are wasted on unnecessary evacuations with attendant loss of business, and the act of crying wolf may lead some to ignore future evacuation orders. Truly, the forecaster's heart knows its own bitterness.

Until the advent of the telegraph in the mid-nineteenth century, the only tools available to the weather forecaster were his own direct visual observations of the sea and sky coupled with his learning and experience. Christopher Columbus evidently learned from the native Caribbean inhabitants that certain progressions of cloud, wind, and sea swell foretold an approaching storm, and he used that knowledge to predict the first hurricane he actually experienced, during his fourth voyage to the New World.

Electronic communications made it possible for the forecaster to assemble simultaneous (or nearly simultaneous) observations covering vast areas, to form a synoptic picture or map of the atmosphere at any given time. By this means, he could discern the patterns formed by wind, temperature, and pressure; and by comparing such patterns at different times, he could detect the movement and evolution of weather systems.

Among the first to do so was Benito Viñes, a Jesuit priest who came to Cuba in 1870 to serve as director of the meteorological observatory of the Royal College of Belen, in Havana. Viñes worked tirelessly to establish a network of meteorological observing stations in Cuba and around the Caribbean, and he issued the first of many hurricane warnings in 1875. He had already made a careful study of the sequence of cloud formations that typically precede hurricanes, and he supplemented this knowledge with information obtained from his network of stations. Viñes and his successors at Belen established Havana at the forefront of the science and art of hurricane forecasting.

During the last half of the nineteenth century and the first half of the twentieth, the number and quality of observations steadily increased, as did both the characterization and the understanding of atmospheric phenomena. Up through World War II, almost all the observations were taken at the earth's surface by dedicated observing stations on land and by ships at sea. A few upper-air observations were obtained by aircraft, kites, and, beginning in the mid-1930s, balloons. The balloon systems, called *radiosondes*, used radio to transmit pressure, temperature, and humidity data back to the ground station. Later, the balloons were tracked by radar to obtain wind information; these systems are called *rawinsondes*. After the end of World War II, rawinsonde networks were rapidly expanded, giving meteorologists their first systematic views of the upper atmosphere and allowing them to incorporate measurements of winds aloft in their forecasts.

Up through the late 1950s, forecasting was mostly a matter of recognizing existing patterns and using both experience and scientific understanding to project those patterns into the future. This was largely a subjective exercise and left much room for error. All that changed when, in 1950, Jule Charney, Ragner Fjørtoft, and John von Neumann at the Advanced Study Institute in Princeton used one of the first digital computers to produce the first numerical weather forecast. In 1957, Akira Kasahara at the University of Chicago carried out the first numerical forecast of the movement of a hurricane. By the 1960s, computer models had become the primary tool of the weather forecaster, and today virtually all forecasts are based on output from such models. But curiously, it was not until the 1990s that the skill of numerical forecasts of hurricane tracks overtook that of simple statistical techniques.

Numerical weather prediction models use computers to solve equations governing the evolution of the atmosphere and, in some cases, of the ocean as well. Called differential equations, they are expressions for the rate of change of various quantities. A simple example of a differential equation is Newton's first law of motion for an object:

$$\frac{dV}{dt} = F \ , \qquad (29.1)$$

which in ordinary English states that "the time rate of change of the velocity equals the force acting on the object." (In this case, the force has been divided by the mass of the object.) The symbol d in the equation stand for "differential." So, to be more precise, the left side of the equation stands for "a small change in velocity divided by the small interval of time over which it occurs." Another word for the time rate of change of velocity is "acceleration."

If F, the force per unit mass, does not vary in time, it is possible to find a solution for the velocity V at any time t:

$$V = V_0 + F \times t,$$

where V_0 is the velocity at time $t = 0$. This says that the velocity at any given time is the velocity at the starting time ($t = 0$) + the force acting on the object multiplied by the amount of time elapsed. We can apply this to the speed of a car, for example. Start from rest ($V_0 = 0$) at time $t = 0$ and depress the accelerator to produce a constant force F. After an amount of time t has elapsed, your speed will be V, as given by the equation I have presented. The more time goes by, the faster you will be going.

This is fine, but in most real problems the force F is not constant, but rather changes with time. If the change is a simple function of time, one can still find a solution to the equation. But if it is a sufficiently complicated function, the solution will either itself be very complicated, or else it may not be possible to write down the solution at all. In such cases, we use computers to solve the equation.

Suppose F varies with time in a complicated but known way. Start with the known value of F at time $t = 0$; let's call that $F(0)$. To make life simple, we will suppose that the velocity at $t = 0$ is also zero. Now choose some time interval and call it Δt. We have to make this as small as possible, but in any case small compared to the time it takes F to change appreciably. So after a small time Δt has elapsed, the velocity will be approximately

$$V(\Delta t) = F(0) \times \Delta t$$

This says that at time $t = \Delta t$, the velocity will be the force evaluated at time $t = 0$ multiplied by the time elapsed, Δt. We have assumed that F has not changed very much over the short time interval. If F is actually constant, then this is an exact solution, but if F varies in time, this will only be approximate. Of course, the smaller we make Δt, the more accurate the solution.

Now we have a (hopefully good) approximation to the velocity at time Δt, so we can reevaluate the force acting on the object. (We are assuming that we know what the force is at all times.) Call the new value of the force $F(\Delta t)$. Now we can take the next step to find an approximate solution of (1) for the velocity at the next time, $2\Delta t$:

$$V(2\Delta t) = V(\Delta t) + F(\Delta t) \times \Delta t.$$

This just says that the new velocity is the old velocity plus the force (updated at the last time) times the amount of time elapsed.

We can keep going indefinitely: Find the small change in velocity that occurs over a small time interval Δt, using the force evaluated at the beginning of the interval, and add it to the old velocity. Now update the force and repeat the procedure.

The smaller Δt, the more accurate the solution, but the longer it will take to calculate it. So, for example, if we want to know the velocity after ten hours, we could repeat the procedure 10 times with a time interval of 1 hour, or 100 times with a time interval of 1/10 of an hour, or 1,000 times with a time interval of 1/100 of an hour. The last of these will be the most accurate, but take longer to get. There is always a trade-off between accuracy and the amount of computation required.

We can do this by hand, with pencil and paper. But it is ever so much faster to use a computer to do it. A procedure like this is called a *numerical integration*.

As a demonstration of numerical integration, we use a set of equations slightly more complicated than equation (29.1). In the first example, we took something that was traveling in a single direction (call it "east"). Now let's consider an object that moves in one or both of two directions, like a car that can be steered. We have to describe the motion in terms of both speed and direction. Alternatively, we can speak in terms of two different speeds: the eastward speed (call it u) and the northward speed (call it v).

Although not a realistic set of equations for the speed of a car, here is a system that is still simple:

$$\frac{du}{dt} = -v, \tag{29.2}$$

$$\frac{dv}{dt} = u. \tag{29.3}$$

Figure 29.1: The eastward velocity as a solution to (29.2) and (29.3). The blue curve represents the exact solution as well as the numerical solution when $\Delta t = 0.01$ hour, and the green curve is the solution using $\Delta t = 0.7$ hour.

▶

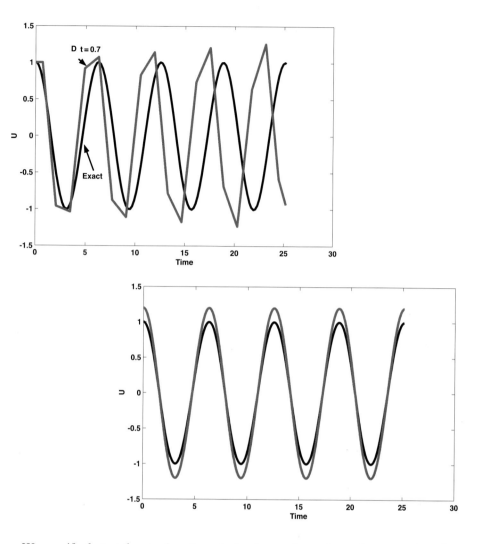

Figure 29.2: The forecast of the eastward velocity u starting from an erroneous initial state is shown by the green curve; the blue curve shows the correct solution.

▶

We specify that at the starting time, $t = 0$ the eastward velocity, u, is 1 mph and the northward velocity, v, is 0.

This pair of equations is coupled, because the forcing for each equation involves the value of the other quantity. As before, we can use the "old" values of u and v to step each equation forward in time, update the values of each, and keep going.

It so happens that this pair of equations has an exact solution, given by

$$u = \cos(t), \quad v = \sin(t), \qquad (29.4)$$

where "cos" stands for "cosine" and "sin" stands for "sine." Figure 29.1 compares this exact solution with the solution obtained by numerical integration using a time step, Δt, of 0.01 hour. The numerical solution is so close to the exact one that it is impossible to see the difference. But if we increase the time step to 0.7 hour, the difference is clear. Although the solution is now inaccurate, it is not wildly so: a short "forecast" would still be useful.

Suppose we had to forecast the speed of this object but we were unable to make a perfect measurement of its current speed (we define "current" to mean "time t = 0"). We estimate the initial value of u as 1.2 mph, whereas its true value is 1 mph. How does this error affect our forecast? The forecast made using this mistaken initial value of u is compared to the actual solution in Figure 29.2. Although the initial mistake leads to an inaccurate forecast, the inaccuracy does not get worse over time. This failure of the initial error to get worse with time is a characteristic of periodic systems like this one. This is why we can make accurate forecasts of, say, the time of sunrise many thousands of years into the future.

The system of equations governing the evolution of the atmosphere is in some ways similar to equations (29.2) and (29.3), but there are more equations and they are more complicated. In addition to the eastward and northward velocity components, there are equations for the upward velocity, temperature, water vapor, cloud water, and precipitation, among other things. To make matters worse, fluids (like the atmosphere) cannot be considered to be a single "object" like a car or a planet. The only real "objects" are the molecules of the gases that make up the atmosphere, but calculating the trajectories of each molecule, even in a cardboard box, would require an inconceivably large computer. To make the problem tractable, we divide the atmosphere into a large but manageable number of pieces. The more pieces, the more accurate the calculation. For each piece, we have to solve the equations for all the quantities mentioned above—velocities, temperature, water vapor, etc.; the equations include the effects of neighboring pieces. So the total number of equations we have to solve is the number of equations for each piece multiplied by the number of pieces. In the case of a contemporary numerical weather prediction model, the number of equations to solve can number in the tens of millions.

The basic equations governing the evolution of fluids were developed during the nineteenth century, and in 1904 the Norwegian physicist Vilhelm Bjerknes suggested that weather could be predicted by solving these equations. In 1922, the British scientist Lewis Fry Richardson published a paper entitled "Weather Prediction by Numerical Process," in which he suggested that the equations be applied to pieces of fluid and integrated numerically, as described earlier. He envisioned a "forecast factory," an arena of 64,000 human calculators organized by a conductor, who would keep the calculation of each piece in sync with the other pieces using flashing lights and telegraphs. In an attempt to prove the viability of the procedure, he spent six weeks doing a single eight-hour numerical weather forecast by hand, only to arrive at an answer that was spectacularly wrong. His forecast was in error not because he had made arithmetic mistakes, but for technical reasons that were not understood until many decades later.

The vision of Bjerknes and Richardson remained in the realm of fancy until the digital computer was invented in the 1940s. Suddenly, what had once seemed an impossible dream became a reasonable expectation. In 1950 the first numerical weather forecast by Charney, Fjørtoft, and von Neumann made use of a drastically simplified set of equations that omitted many phenomena but avoided the technical problems that had plagued Richardson's attempt.

The 1950s were a time of enormous optimism for weather forecasters. Digital computers were getting faster and more powerful by leaps and bounds, enabling the atmosphere to be divided into ever finer pieces and the equations to include more and more physical processes. At the same time, observations of the atmosphere were improving, providing an increasingly accurate specification of its state for numerical forecasts. Given the great successes of numerical integration in other realms, such as calculating the trajectories of planetary orbits and long-range shells fired by guns, there was every reason to believe that weather forecasts would continue to improve indefinitely.

This spirit of optimism was forever shattered one winter day in 1961, in a nondescript office at the Massachusetts Institute of Technology. Edward N. Lorenz, a professor of meteorology, was experimenting with the numerical integration of some equations not much more complicated than (29.2) and (29.3):

$$\frac{du}{dt} = s \times (v - u),$$

$$\frac{dv}{dt} = r \times u - v - u \times w,$$

$$\frac{dw}{dt} = u \times v - b \times w.$$

Here s, r, and b are just constant numbers. These equations were meant to approximate the behavior of convection in a fluid heated from below and cooled from above.

For certain values of the parameters s, r, and b, the solutions of these equations are very much like the solutions shown in Figure 29.1; they just vary periodically in time. For other combinations, the solutions are more complex. Lorenz was examining one such solution and decided he would do the integration over again, because he wanted to integrate it out further in time. He copied down the same numbers he had used the first time and started it over. After stepping out to get a cup of coffee, he noticed that the computer was producing a completely different solution. At first he assumed that one of the many vacuum tubes in the primitive machine had malfunctioned, a common occurrence. He summoned technicians, who pronounced it healthy.

He then rechecked whether he had typed in the same numbers as were used in the first integration. All he had done was to round off the six digits of the computer printout to three (for example, using .832 instead of .832479), introducing an error of one in a thousand.

Finally he realized that this was the problem.

Figure 29.3 shows two solutions of Lorenz's equations, one starting from a certain initial state and the second starting from a state differing from the first by one part in a thousand. At first, the two solutions are indistinguishable. Then, quite suddenly, they diverge to very different solutions.

Apparently, in certain systems of equations, very slight errors in the specification of the initial state of the system can lead to enormous errors in the forecast, quite unlike simpler systems, such as the periodic system described earlier in this chapter, in which small initial errors lead to correspondingly small forecast errors. This finding was foreshadowed 55 years earlier in an essay entitled "Science and Method," written by the French mathematician Henri Poincaré:

> If we knew exactly the laws of nature and the state of the universe at the initial moment, we could predict exactly the state of that same universe at a succeeding moment. But even if the natural laws no longer held any secrets for us, we could still only know the initial state *approximately*. If that enabled us to predict the succeeding state with the same approximation, that is all we require, and we would say that the phenomenon had been predicted, that it is governed by laws. But it is not always so; it may happen that *small differences in the initial conditions produce very great ones in the final state*. A small error in the former will engender an enormous error in the latter. Prediction

Figure 29.3: Two solutions of the Lorenz equations for the variable u as a function of time. The second solution begins from a state that differs from that used in the first solution by one part in a thousand.

▶

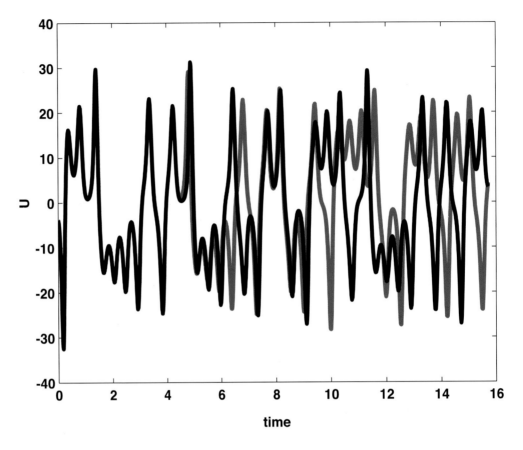

becomes impossible, and we have the chance phenomenon. [Emphases as in the original.]

Lorenz's confirmation of Poincaré's conjecture shattered the illusion that we can continue to improve weather forecasts indefinitely, simply by improving the equations and observations of the atmosphere. Experiments with sophisticated weather prediction models, having millions of equations, show the same behavior as Lorenz's simple model: very small differences in the initial state eventually amplify to render the forecasts useless. These same experiments show that even with excellent and very extensive measurements of the atmosphere, and very good models, it will be impossible to make detailed weather predictions beyond about two weeks. (This does not necessarily preclude predicting changes in weather patterns. For example, we might someday be able to predict whether next fall will be, on average, colder than normal. There has been some success in predicting El Niño/La Niña six months or so ahead of time. This might lead to an expectation that rainfall in southern California will be greater than average during the coming winter, or that there will be more than the usual number of hurricanes in the Atlantic next summer.)

In spite of the pessimistic conclusions of chaos theory, as Lorenz's work has come to be known, numerical weather forecasts improved dramatically during the 1960s, '70s, and '80s. Unfortunately, improvement slowed nearly to a standstill beginning around 1990, perhaps reflecting the ultimate limits predicted by chaos theory.

Today, virtually every weather forecast is based on the output of numerical weather prediction models, which are run on some of the world's largest supercomputers. Improvements in numerical weather forecasts rely on four major efforts:

1 ► *More and better measurements of the atmosphere.* While the number of conventional observations, such as from surface stations, balloon soundings, ships, and buoys, have actually decreased in recent years, there has been a dramatic increase in the amount and usefulness of satellite data and measurements made from commercial airliners. There is ongoing debate about which combinations of platforms and devices offer the most cost-effective means of measuring the atmosphere for the purpose of making weather forecasts, and while satellites are an essential source of data, the lowly and comparatively cheap weather balloon still provides many of the useful measurements used in numerical weather forecasts (except in the Southern Hemisphere, where there are no longer enough weather balloons to make much difference).

2 ► *Improved ways of using atmospheric measurements to initialize numerical forecasts.* Initializing a numerical forecast is a far more complex problem than, say, initializing a forecast of the orbit of a planet. In practice, one begins with a first guess about the current state of the atmosphere. This is usually a 12-hour-old forecast of the current conditions, and it represents the best guess of the actual state absent any measurements. Observations are then used to modify this first guess to bring it more in line with reality, in a process called *data assimilation.* This is a very active area of research, and it is contributing to substantial improvements in

the specification of the initial state of the atmosphere.

3 ► *Better accuracy of numerical algorithms.* As computers become faster, it is possible to divide the atmosphere into an ever-larger number of smaller and smaller pieces while still being able to complete the forecast reasonably quickly. This, together with better ways of doing numerical integration, is making the integrations themselves more and more accurate.

4 ► *Improved model physics.* Although the equations themselves have been known for more than a century, there are many physical processes that occur on space and time scales that are still too small for models to explicitly resolve. Small-scale turbulence, for example, is important for moving heat, moisture, and momentum from one place to another, particularly in the layer of air next to the earth's surface, but the eddies themselves are too small to resolve. Yet their effects cannot be ignored, and so they are represented in terms of quantities the model can resolve. For example, the rate at which turbulence carries heat and moisture from the ocean to the interior of the atmosphere is represented as a function of the average (resolved) wind speed near the surface and a few other variables. These representations, often called parameterizations, are by their very nature approximate, and their development is something of an art form. Improved measurements of the atmosphere do allow these representations to be more rigorously tested, however.

Although work in these four areas will certainly lead to further improvements, chaos theory tells us that such improvements will be more and more incremental in nature. Uncertainty will always be with us, and will be occasionally large. For example, a hurricane whose track takes it very near a "saddle point" in the flow (see

Figure 29.4: Eight forecasts of the movement of a hurricane, using the same model but starting from eight slightly different estimates of its initial location, at lower right.

▶

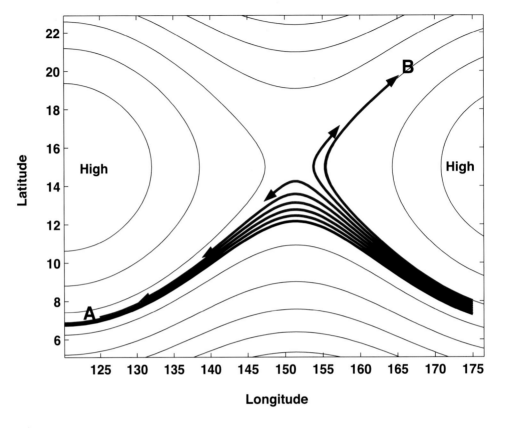

Chapter 18) may end up in the Gulf of Mexico or in the central North Atlantic, depending on very small differences in its current position.

The realization that we can never rid ourselves of uncertainty led atmospheric scientists to develop ways of estimating how much uncertainty is associated with each forecast. There are some meteorological circumstances in which the uncertainty in, say, a three-day forecast is small and others in which it is large. We want to be able to forecast just how large or small the uncertainty itself is. One way to do this is to make not just one numerical forecast, but many, all beginning from slightly different estimates of the initial state of the atmosphere and/or using slightly different model physics. This technique is called ensemble forecasting.

As a very simple example of ensemble forecasting, consider a hurricane embedded in a flow that does not evolve with time but that contains a saddle point. Because the initial position of the hurricane is uncertain, we run eight forecasts using the same numerical model with eight slightly different but equally probable estimates of the initial position of the storm. The result is shown in Figure 29.4. Even though the initial positions of the storms are within 20 mi of each other, the final positions after six days are thousands of miles apart, an example of large sensitivity to initial state. But once the storm "turns the bend" in the middle of the figure, there is very little uncertainty in where it will go; in fact, the uncertainty in the path it will take then decreases with time, although the time it takes to reach a given point along that path still varies quite a bit from one ensemble member to the next. A forecaster confronted with the situation could state that, with high probability, the hurricane will hit

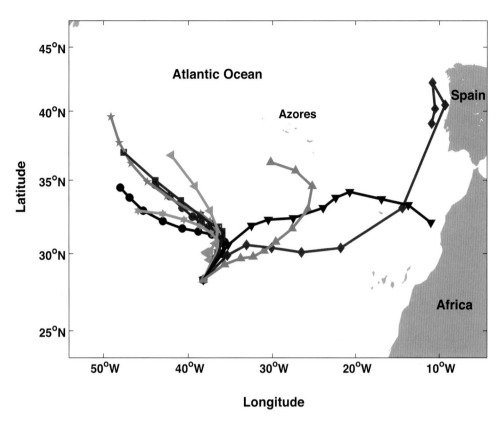

Figure 29.5: Numerical forecasts of the track of Hurricane Kate in 2003. Each track represents the prediction of a different numerical model, starting from Kate's observed location over the central North Atlantic.

▶

either point A or point B, but in either case, the time it would hit one of these points is uncertain.

This is an example of "Lagrangian" chaos: the flow itself is exactly predictable in this case (it is not changing in time), but there is uncertainty in the trajectory of a particle (or storm) that is carried along by the flow. Unfortunately, the real world is even more complicated, because the large-scale airflow does evolve in time in a way that is itself chaotic, so we often have to deal with chaotic trajectories in a flow that is also chaotic.

A real example of an ensemble hurricane forecast is presented in Figure 29.5. In this case, the initial position is the same for each ensemble member, but different numerical models were used to generate each forecast. All but one have the storm traveling northward during the first three days of the forecast (which, as it turns out, the real storm did), but then they diverge greatly.

Thus the forecaster can have some confidence that the hurricane will travel northeastward from its current location in the Atlantic southwest of the Azores, but then he would attach great uncertainty to the forecast beyond three days.

The widespread use of multimodel ensembles has another salutary effect on forecasting: by scrutinizing different models on a regular basis, forecasters learn which ones are more dependable, and this creates a keen sense of competition among model developers, leading to faster model improvement. Weighted averages of the various model forecasts are also proving useful.

Ensemble forecasting is yielding rapid progress both in the overall skill of forecasts and in the quantification of forecast uncertainty. Weather-sensitive enterprises, like aviation and agriculture, are just beginning to learn how to use forecasts of uncertainty.

In addition to technical advances in hurricane forecasting, there has been considerable evolution in the response of markets to forecasts. For example, private forecasting operations have recently undertaken "meta-forecasting," which quite literally forecasts what the official forecast will say. For example, if the official forecast calls for a hurricane to affect Charleston, South Carolina, then grocery and hardware stores in and around Charleston will experience a run on certain items such as flashlights and bottled water. This happens regardless of whether the hurricane ends up affecting Charleston. If the official forecast can be anticipated, savvy suppliers can rush these items to market ahead of their competition.

Of somewhat more consequence is the recent invention of the "catastrophe bond," or "CAT-bond," for short. These bonds help insurance companies capitalize rare but extremely expensive disasters, like Hurricane Andrew of 1992. They return principal and interest—often at large rates—like debt securities do, as long as the insurer does not suffer a loss beyond some agreed-upon threshold. For example, in June 1998, one insurance company floated a $477 million CAT bond with a hurricane loss threshold of $1 billion. Because that threshold was not exceeded, investors enjoyed the return of their principal plus 11 percent interest. Had it been exceeded, they stood to lose everything. Though most CAT bonds cover risks over a hurricane season, it is not inconceivable that similar bonds will be traded over individual storms, greatly increasing the financial consequences of forecasts. One can only guess whether these developments will contribute more to the joy or to the bitterness of the forecaster's heart.

▲

Hurricane 1999. Oil
painting by Haitian artist
Vilaire Rigaud, 1999.

H igh noon I can't believe my eyes
Wind is ragin' there's a fire in the sky
Ground shakin' everything comin' loose
Run like a coward but it ain't no use
Edge of the river just an ugly scene
People getting pushed, and people gettin' mean
A change is comin; and it's gettin' kind of late
There ain't no survivin', there ain't no escape

—"Change in the Weather"
a song by John Fogerty (b. 1946)

30　　Cyclone Tracy

Christmas 1974 divided Darwin time into BT (before Tracy) and AT (after Tracy).

　　—Peter Copland

On early Christmas morning 1974, the city of Darwin in Australia's Northern Territory was razed by one of the strongest tropical cyclones ever recorded on that continent. The storm killed 65 people and injured another 790 while destroying more than 80 percent of Darwin's buildings and uprooting every tree in the city. A storm surge of 12 ft inundated the coast and capsized 21 vessels. Tracy caused at least $800 million in damages, making it the most costly tropical cyclone in Australia's history.

　　Situated in far northwestern Australia, Darwin is highly susceptible to strong cyclones. In 1875 the ship *Gothenburg*, en route from Darwin to Australia's east coast, was sunk by a cyclone, killing 102, almost a quarter of Darwin's fledgling population. Destructive storms struck the city again in 1878, 1881, 1897, 1917, and 1937. (Adding to its misery, Darwin was bombed about 60 times during World War II. The city's general misfortune was attributed by the region's Larrakia Aboriginal people to their pagan god Nungalinya, who became enraged by white settlers.)

　　The storm that laid waste to the city seems to have been tailor-made for destruction. Developing early on December 20 over the western Arafura Sea, about 500 km (300 mi) northeast of Darwin, Tracy moved southwestward, becoming a tropical storm on the twenty-first (Figure 30.1). By the afternoon of the twenty-second, maximum winds had increased to over 45 m/s (100 mph), and the storm was clearly visible on Darwin's meteorological radar. During the early morning of the twenty-third, Tracy's forward movement slowed to a mere 2 m/s (4 mph) as it passed just west of Bathurst Island, off Australia's northwest coast. From this time on, Tracy moved over water no deeper than 50 m (160 ft). Under such conditions, the warm layer of water at the ocean surface extends right down to the bottom, so there is no cold water to upwell to the surface and reduce the strength of the storm (see Chapter 12).

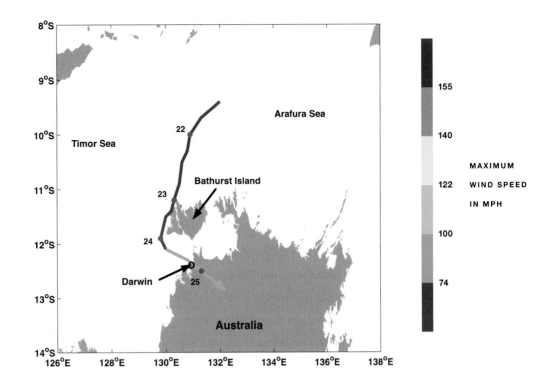

Figure 30.1: Track of Cyclone Tracy of 1974. Numbers show dates in December.

▶

As Tracy moved slowly over the shallow coastal waters of the Timor Sea, it quickly strengthened, attaining winds of over 55 m/s (125 mph) late on December 23. By the morning of the twenty-fourth, forecasters at the Darwin Tropical Cyclone Warning Center were gravely concerned, predicting that highly destructive winds would occur later that day on Bathurst Island. They expected Tracy to be located 100 km (60 mi) west of Darwin by 9 P.M. that evening. Shortly after this statement was issued, Tracy began a fateful turn toward the east, on a glide path to Darwin. By afternoon, the storm

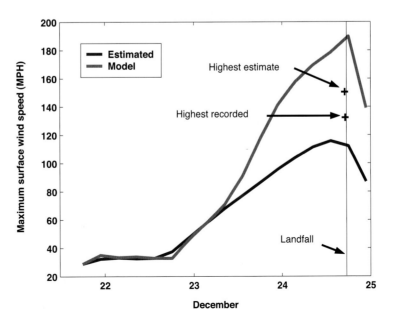

Figure 30.2: Maximum wind speed in Tropical Cyclone Tracy. The blue curve shows the official estimate, and the green curve shows the results of a computer simulation. The highest recorded wind speed and the highest estimated wind speed are also shown.

▶

was just 80 km (50 mi) west of Darwin, with maximum winds exceeding 65 m/s (150 mph). A cyclone warning issued at 10 P.M. local time stated that severe tropical cyclone Tracy was centered 41 km (25 mi) west-northwest of Darwin and moving east-southeast at 2 m/s (4 mph). Forecasters expected the center to cross the coast in the vicinity of Darwin in the early hours of the next morning.

Tracy was a remarkably small storm. Its eye was estimated to have been only 12 km (7 mi) in diameter when the storm made landfall early on Christmas morning, making the eye the smallest on record anywhere in the world. With an eye narrower than its depth of around 15 km (9 mi), the storm resembled a large funnel, intermediate in size between a large tornado and an average hurricane. Gale force winds extended only 40 km (25 mi) from the storm center, so that even late on Christmas Eve, the winds had not yet reached gale force in Darwin. Although forecasts at this time clearly warned of a direct hit, it must have been hard to take them seriously, given the relatively weak winds just 40 km (25 mi) from the center. The tiny size of the storm also made it difficult to estimate its strength, given the complete absence of measurements near its core.

Tracy crept inexorably toward the city. Around midnight, Christmas Eve, winds reached gale force and accelerated to a terrific peak sometime after 3 A.M. The last warnings were issued at 2:30 A.M., shortly before both of Darwin's radio transmitters failed. At 3:05 A.M. a peak wind of 60 m/s (135 mph) was recorded at Darwin Airport, just as the anemometer blew away. The eye passed over the airport between 3:50 and 4:25 A.M., after which extreme winds returned. By 7 A.M. the storm was over in Darwin.

The history of Tracy's intensity is not well known, even at landfall, owing to its tiny size and immense destructiveness. A computer simulation of the storm (Figure 30.2) yields a peak wind of around 85 m/s (190 mph) at landfall, one of the highest wind speeds ever produced by the model used. These extreme wind speeds were reached in the simulation because of an optimum combination of circumstances. First, the Timor Sea, one of the warmest bodies of water in the world, was close to its seasonal peak temperature. As mentioned earlier, Tracy happened to move over shallow coastal waters, so that after the morning of December 23, no further upwelling of cold water occurred. Finally, Tracy's small size and unusually slow forward movement allowed it enough time to strengthen over the hot, shallow coastal seas before striking Darwin. To top off this fateful combination of factors, Tracy passed directly over the city, and because of its tiny diameter, winds accelerated from a gentle breeze to inconceivable fury in a scant five hours, giving residents little forewarning.

While we will never know just how strong Tracy was, we do know that every single tree in Darwin was uprooted and stripped of its foliage, testimony to an exceptionally violent storm. In 1974, Darwin still had the character of a frontier town, with many of its buildings constructed of corrugated iron. These proved no match for Tracy, and the devastation was almost total (Figure 30.3). Many victims were impaled by flying timbers or sliced by sailing sheets of tin roofing from the disintegrating government-built housing. As the storm waned, looters took to the streets, carting off whatever they could pilfer from homes and businesses, only to discover that their own homes had blown away.

For ten hours after the storm, there was no way to tell the world that Darwin had been razed. When news of the disaster finally reached Australia's capital, Canberra, the government was faced

Figure 30.3: Darwin in the aftermath of Cyclone Tracy.

▶

with 36,000 homeless people needing food and shelter. Over the next six days, 35,000 people left Darwin in an exodus the likes of which had not been seen since the bombing runs of World War II.

Darwin has been rebuilt, stronger this time, and is today a thriving and lively city. History instructs, though, that it is only a matter of time—probably a few decades or less—before disaster once again rolls in from the sea.

▲

After the Hurricane.
Watercolor by American
artist Ogden Milton
Pleissner.

To whichever point of the compass the eye was directed, a grand but distressing ruin presented itself. The whole face of the country was laid waste; no sign of vegetation was apparent, except here and there small patches of a sickly green. The surface of the ground appeared as if fire had run through the land, scorching and burning up the production of the earth. The few remaining trees, stripped of the bows and foliage, wore a cold and wintry aspect.

—William Reid (1791–1858), *An Attempt to Develop the Laws of Storms,* 1838

31 Hurricane Andrew, 1992

*More than once, a society has been seen to give way before the wind which is let loose upon mankind;
history is full of the shipwrecks of nations and empires; manners, customs, laws, religions—and
some fine day that unknown force, the hurricane, passes by and bears them all away.*

—Victor Hugo, *Les Miserables*

*There's still a tremendous complacency. People hear me tell my story and they do nothing about it.
I'm telling you, we are sitting ducks.*

—Stanley Goldenberg, hurricane researcher, after barely surviving Andrew

In the forty years from 1926 to 1966, Dade County, Florida, where Miami is located, was struck by
hurricanes 13 times. But there were no hurricanes at all for the next 25 years, during which the
county's population more than doubled. This lull in hurricane activity proved to be the proverbial
calm before the storm, writ large, and it induced a sense of complacency among all but the oldest residents. While building codes actually became more stringent during this lull, enforcement waned, and
by the late 1980s, there were only 16 building inspectors for a population of well over 1 million. It was
as though the storm gods had primed south Florida for catastrophe.

That catastrophe was Hurricane Andrew, the costliest natural disaster in U.S. history and the
third most intense hurricane ever to strike the country. Remarkably, only 26 died in Florida as a direct
result of the storm, but tens of thousands were left homeless, and at least 40 more died in the aftermath. Andrew changed the face of south Florida, and brought about sea changes in the insurance and
construction industries.

On August 14, 1992, Andrew formed, as many Atlantic hurricanes do, from a tropical wave that
moved offshore from Africa. A little better organized than most, the wave moved westward at about 25
mph. At 18:00 UT on the sixteenth, the National Hurricane Center declared that a tropical depression had formed, and by 12:00 UT on the seventeenth, while moving westward through the central
tropical North Atlantic, Andrew became the first tropical storm of the season.

Over the next three days, the storm moved northwestward around the Bermuda high-pressure
system, a persistent feature of the summertime flow over the North Atlantic. While over the Sargasso
Sea, it encountered increasing southwesterly winds aloft, associated with an upper-level disturbance
moving eastward off Florida. Andrew fought the wind shear, producing only sporadic convection for

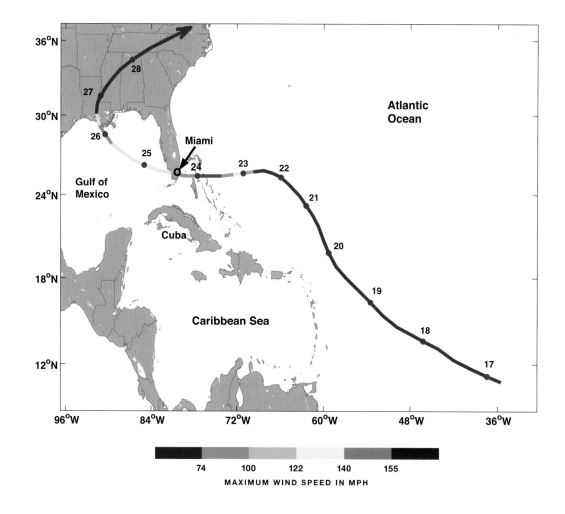

several days. On the morning of the twentieth, an Air Force reconnaissance mission found only a weak, diffuse, low-level circulation, leading forecasters to predict that "the struggle may be ending soon." Had the upper-level trough been just a little stronger, or Andrew a little weaker, south Florida might have been spared disaster. But Andrew proved tenacious: though it lacked a well-defined low-level circulation, winds near hurricane force were recorded only 1,500 ft above the surface.

During the night of the twentieth and morning of the twenty-first, the upper-air trough that had nearly destroyed Andrew moved out to the east. This had two major effects: the strong southwesterly winds aloft that had steered Andrew northward collapsed, so that the storm now moved more westward with the low-level trade winds. The departure of the trough also removed the wind shear just as Andrew was passing over extremely warm water. The effect was like removing the lid from a boiling cauldron: Andrew exploded, reaching hurricane intensity by the morning of the twenty-second and achieving its astounding peak intensity of 170 mph late on the twenty-third. Although forecasters had anticipated some intensification, its ferocity caught them off guard. So too did its forward movement. On Friday, August 21, the computer models on which forecasters rely showed Andrew slowing down, with only a 7 percent chance of affecting Miami by 2 P.M. the following Monday. Many south Florida residents heard a broadcast from the National Hurricane Center on Friday wishing them a good weekend and encouraging them to check the reports again on Sunday or Monday. But by Saturday after-

Figure 31.2: Andrew taking aim at south Florida.

▶

noon, August 22, forecasters knew they were dealing with a killer, bearing straight down on south Florida at an increasing pace. On Sunday morning, all but the most out-of-touch residents knew that a major storm was imminent. The question was where, precisely, would it strike? Floridians prepared for the onslaught.

Andrew was an extremely intense hurricane as it passed over Eleuthera Island and the southern Berry Islands in the Bahamas late on the twenty-third, but along the way it lost some strength, with maximum winds diminishing from their peak of 170 mph to a "lull" of 145 mph. But Andrew then passed over the very warm, deep waters of the Gulf Stream, and by the time it made landfall near Homestead in the predawn hours of Monday, August 24, it was once again intensifying rapidly. So inexorable was Andrew's quest to regain its former, staggering intensity that it continued to intensify even *after* landfall, according to measurements made by reconnaissance aircraft.

No one is quite sure just how strong Andrew was when it hit south Florida, because most of the instruments capable of measuring wind and pressure were destroyed before the storm reached its peak. (Among the casualties was the radar atop the National Hurricane Center in Coral Gables, which blew over into an adjoining residence shortly after recording a truly frightening image of Andrew bearing down on south Florida.) Initially, the maximum winds at landfall were reported to be 145 mph, but a panel convened in 2002 to assess Andrew's intensity concluded, based mostly on engineering analyses of the damage, that the highest sustained winds were about 165 mph. The revised history of Andrew's maximum wind speeds, together with the results of a computer simulation, is shown in

Figure 31.3: History of the maximum wind speed in Hurricane Andrew. Observations are shown by the blue curve; the computer simulation by a green curve.

▶

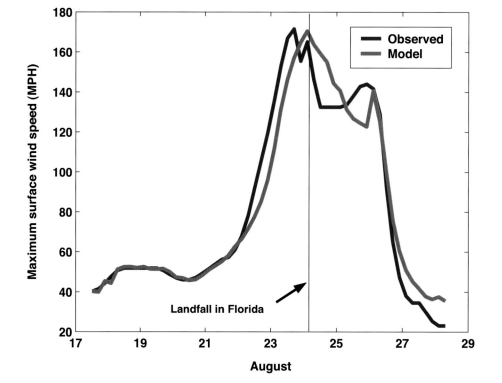

Figure 31.3. The model fails to achieve Andrew's first climax over the Bahamas, but it gets the landfall intensity nearly right.

While there may be doubt about Andrew's exact intensity at landfall, there is no doubt about the outcome. In just a few hours on that early Monday morning, more property was destroyed than in any single natural event in U.S. history. Although an impressive storm surge of nearly 17 ft inundated the coast south of Miami, Andrew's fierce winds inflicted most of the damage, tearing houses apart, lifting cars off the street, and uprooting trees. Many residents had taken the advised precautions, such as boarding up windows, but these proved no match for the violent winds near the storm center. While everyone in the path of the storm supposed that they were in for a blow, no one imagined they would be crouching in mortal fear in interior hallways (few Florida houses have basements) while their world collapsed around them, accompanied by horrifying sounds likened to freight trains and screeching banshees. Houses and other buildings were smashed to pieces or severely damaged by falling trees, by ballistic debris from other structures disintegrating upwind, or by the sheer, unrelenting force of the wind. Through the dark tumult, only the lone voice of a forecaster, Brian Norcross, penetrated the stormy cacophony and the static of battery-powered radios to give comfort in the night. Yet a great many residents doubted they would survive to see the next day. In view of the nearly total destruction of huge swaths of real estate, it is a miracle that so many did.

As daylight slowly returned and the wind eased during the morning of Tuesday, August 25, survivors emerged, stunned, from the debris. Some wept, some were stoic, and many were so dazed they did not recognize their profoundly altered surroundings. In many places, little but rubble stretched as far as the eye could see. What few trees remained standing were completely stripped of

Figure 31.4: Aerial view of Andrew's effect on a south Florida neighborhood.

▶

foliage, painting a bleak, wintry scene at stark odds with the searing heat of the storm's aftermath. There was no power, no water, and no vehicles on the roads, and a profound silence replaced the howling of the previous night. For most, there was no place to go and no way to get there, with fallen trees and other debris blocking almost every road. Many survivors carry with them to this day the psychological detritus of trauma, panicking at the sound of thunder or the appearance on satellite maps of suspicious cloud clusters.

Help was slow to arrive. Many hospitals, fire stations, and other emergency facilities near the storm's path were themselves incapacitated, and the destruction of most forms of communication slowed the flow of news to the outside world. When, after several hours, the scope of the disaster began to be appreciated, relief efforts finally got underway, but they were hampered by a lack of coordination and leadership. Although it was clear almost immediately that the magnitude of the tragedy completely overwhelmed local and state resources, the federal government was slow to respond, prompting the Dade County emergency manager to cover the front page of the *Miami Herald* with a cry for help. Although the federal government commanded tremendous resources, it, too, suffered a lack of coordination, and it took many weeks to ramp up relief efforts.

The extent of the damage boggles the mind. The storm completely destroyed about 63,000 of Dade County's 528,000 residences and damaged another 110,000. Nine public schools were reduced to rubble and another 23 heavily damaged. More than 9 out of every 10 mobile homes were obliterated in south Dade County, and only 1 percent of the city of Homestead's mobile homes survived intact. The storm virtually obliterated south Florida's lime, avocado, tropical fruit, and nursery industries. All told, Andrew caused more than $30 billion in damages in Florida, and another

$1 billion in Louisiana, severely straining the U.S. insurance industry, with nine insurers declaring insolvency. It is sobering to reflect that Andrew missed the economic core of south Florida, passing some tens of miles south of downtown Miami and sparing most of the city. Had it tracked just slightly north, the death toll and damage would surely have been far worse.

Detailed analysis of the damage wrought by Andrew revealed disturbing patterns. Houses built during the boom years after 1980 were more severely damaged than older structures. Quite a few aerial photos show largely intact, stately houses built in the early twentieth century surrounded by the wrecked remains of the expensive but flimsy products of a more recent real estate gold rush. To many, this came as a surprise, since Dade County has among the toughest building codes in the country. An investigation by the *Miami Herald* pointed the finger: much of the damage could be attributed to poor compliance with the building codes, which in turn could be traced to the corrupting influence of construction businesses on building inspection. At the height of the building boom, the construction industry accounted for fully a third of all the funds contributed to election campaigns for the Metro Commission, the agency that interprets and oversees the building code for south Florida. In the years leading up to Andrew, the majority of the Commission's members came from the construction industry. Not surprisingly, the inspection process broke down, with fewer and fewer inspectors for more and more buildings. By the late 1980s, so few inspectors were on the job that each one conducted as many as 70 inspections per day, a rate of one every six minutes, not accounting for time in transit. Some inspectors conducted their inspections from their cars.

It is no surprise that many houses were simply not built to code.

No natural catastrophe is entirely natural. For a hurricane to be regarded as a catastrophe at all, we human beings and our belongings must be in harm's way. The ecology of Florida and other hurricane-prone places has evolved in concert with hurricanes, which are part of the course of nature; the Everglades suffered little from Andrew and rebounded quickly. The ancient Mayans learned to build their cities away from the coast, an adaptation that largely spared them the ravages of hurricanes. One might have thought that we, too, would adapt to experience and learn to deal with a fact of nature. Yet we seem content to suffer again and again the same calamity. Indeed, the U.S. government's flood insurance program forces farmers in Kansas to subsidize the risk wealthy Floridians take in building oceanfront houses highly vulnerable to storm surges. For decades before Andrew, successive directors of the National Hurricane Center and others involved in emergency preparedness traveled around the country, from the smallest coastal villages to the halls of Congress, warning that our habit of building flimsy structures in hurricane-prone places was setting us up for disaster.

Did Andrew, finally, teach us a lesson? The answer, if one is to believe Stanley Goldenberg, quoted at the beginning of this chapter, is apparently no.

▲

After the Hurricane, Bahamas. Watercolor by American artist Winslow Homer, 1899.

D ay breaks through the flying wrack, over the infinite heaving of the sea, over the low land made vast with desolation. It is a spectral dawn: a wan light, like the light of a dying sun.

The wind has waned and veered; the flood sinks slowly back to its abysses—abandoning its plunder,—scattering its pitieous waifs over bar and dune, over shoal and marsh, among the silences of the mango swamps, over the long low reaches of sand-grasses and drowned weeds, for more than a hundred miles. From the shell-reefs of Pointe-au-Fer to the shallows of Pelto Bay the dead lie mingled with the high-heaped drift....

—Lafcadio Hearn (1850–1904), *Chita: A Memory of Last Island*

32   Hurricanes and Climate

Another problem, of much more far-reaching consequences, presents itself. What kind of secular changes may have existed in the frequency and intensity of the hurricane vortices of the Earth? And what changes may be expected in the future? We know nothing about these things, but I hope [to] have shown that even quite a small change in the different factors controlling the life history of a hurricane may produce, or may have produced, great changes in the paths of hurricanes and in their frequency and intensity. A minor alteration of the surface temperature of the sun, in the general composition of the earth's atmosphere, or in the rotation of the earth, might be able to change considerably the energy balance and the balance of forces within such a delicate mechanism as the tropical hurricane. During certain geological epochs, hurricanes may have been just as frequent as the cyclones of our latitudes, or they may have occurred all over the oceans and within all coastal regions, and they may have been even more violent than nowadays. During other periods they may have been lacking altogether. In studying paleo-climate and paleo-biological phenomena, especially along the coasts of previous geological epochs, it may be wise to consider such possibilities.

—Tor Bergeron, 1954

Two Louisiana State University graduate students balance precariously in a rubber dinghy, trying to lower an ungainly piece of equipment to the bottom of a marshy lake in coastal Alabama. They and their supervisor, Professor Kam-Biu Liu, are using the equipment to obtain long, cylindrical cores of sediments from the bottom of the lake. Back at the lab, their hopes are confirmed: interspersed among the thick layers of black, organic muck laid down when dead plants settled to the lake bottom are several thin layers of sand (Figure 32.1). Professor Liu and his student, Miriam Fearn, had hypothesized that when violent hurricanes make landfall near this lake, storm surges wash over the dunes separating the lake from the Gulf of Mexico, depositing sand in the lake bottom.

Liu's team carefully sections the core, sending the black organic layers to a laboratory for radiocarbon dating. High in the atmosphere, cosmic rays collide with isotopes of nitrogen and form an isotope of carbon, ^{14}C, which rapidly oxidizes into an isotope of carbon dioxide, $^{14}CO_2$. This gas makes its way down to the surface, where it is incorporated into plants in proportion to the amount in the atmosphere. But when the plants die, this carbon isotope is no longer exchanged with the atmosphere and decays radioactively. The radiocarbon lab carefully measures how much ^{14}C remains in the organic

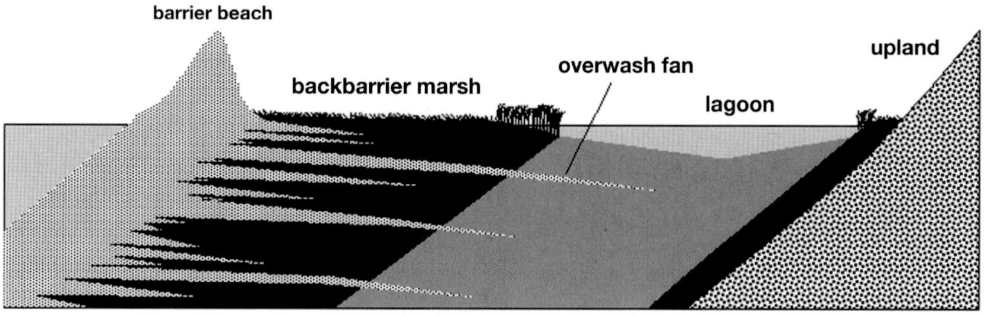

Figure 32.1: Storm surges wash sand from a barrier beach (left) into marshes and lagoons. With time, the sand is buried under dead plants (black material), forming interleaving layers of sand and mud. Cores drilled from the lagoon or marsh reveal the mud and sand layers.

▶

barrier beach

backbarrier marsh

overwash fan

upland

lagoon

matter in the cores and thereby estimates how much time has passed since the plant matter was deposited on the lake bottom. By dating the organic ooze on either side of each sand layer, Liu's team can make an accurate estimate of when the sand was deposited. The core Liu is working on has sand layers dating back more than three thousand years.

Further north, along the coastal plains of Connecticut and Rhode Island, Drs. Jeffrey Donnelly and Thompson Webb are taking cores from a marsh just behind a sand barrier separating it from Long Island Sound (Figure 32.2). The sand layers they find record not only some of

New England's great historical hurricanes, but also prehistoric storms dating to the early fifteenth century.

Liu, Fearn, Donnelly, and Webb, and their associates are pioneers of the new field of paleotempestology, which seeks to create a long-term record of hurricane activity through examination of the geological record. They and other researchers hope to be able assess how the frequency and intensity of hurricanes may have varied in the past, fulfilling the hope expressed long ago by the Norwegian meteorologist Tor Bergeron in the epigraph of this chapter. By understanding how hurricanes may have varied

Figure 32.2: Examples of sand layers in a marsh at Barn Island, Connecticut, deposited during the Great New England Hurricane of 1938 and Hurricane Carol of 1954.

▶

1954 (Carol)

1938

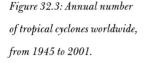
Figure 32.3: Annual number of tropical cyclones worldwide, from 1945 to 2001.

▶

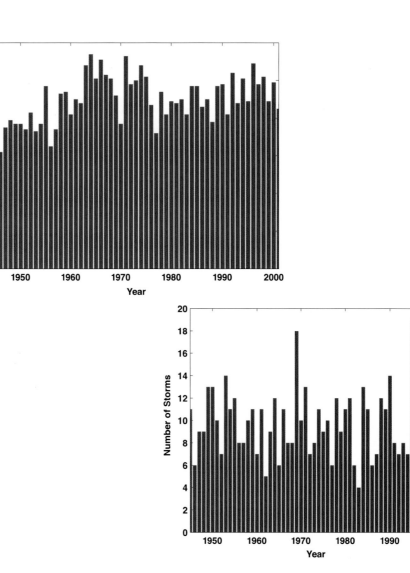

Figure 32.4: Annual number of tropical cyclones over the North Atlantic, from 1945 to 2001.

▶

with known changes in the earth's climate, we may be able to answer such thorny questions as how global warming might affect hurricane activity.

While paleotempestology may reveal variations of hurricane activity on time scales of hundreds to thousands of years, the historical record allows us to look at fluctuations that occur from year to year and decade to decade. Reliable records of landfalling hurricanes date back a century or more along the U.S. East Coast, and village records in coastal China note typhoons as early as 500 A.D. Unfortunately, even such long records do not contain enough storms to draw definitive conclusions about changes in activity,

except in the limited coastal regions to which they pertain.

Since about 1950, aerial reconnaissance and improved observing techniques have provided a fairly comprehensive record of tropical cyclones the world over. Even so, some storms that spent their lives at sea far from shipping lanes no doubt avoided detection. It was not until the satellite era began in the 1960s that virtually all tropical cyclones were routinely observed.

Figure 32.3 shows the annual number of tropical cyclones worldwide since 1945. The frequency of storms appears to increase from the late 1950s to the mid-1960s, but this very likely reflects increasing detection by satellites during this period. Since the mid-1960s, the number of storms has remained fairly stable, ranging from about 80 to about 100 per year, with an average of 90. Scientists would like to know what controls this number. Why aren't there more or fewer storms?

Although the global annual tropical cyclone count is fairly stable, it varies a great deal from one year to the next in individual oceans. Figure 32.4 shows the annual number of tropical cyclones over the North Atlantic (including the Caribbean and the Gulf of Mexico). Here the number of storms varies greatly, from a minimum of 4 in 1983 to a maximum of 19 in 1994. Are these variations random, or are they related to other global climate variations? William Gray, Christopher Landsea, and their colleagues at

Colorado State University showed that Atlantic storm activity is profoundly influenced by a variety of climate phenomena, especially El Niño.

Long ago, Peruvian fishermen noticed that every few years, the water in which they fished warmed up rather quickly, greatly reducing their catch. Perhaps somewhat cynically, they called this phenomenon El Niño, "the Christ child," because it often happens around Christmas. Measurements from satellites, ships, and buoys reveal El Niño to be a complex phenomenon that affects ocean temperatures across virtually the entire tropical Pacific, also affecting weather in other parts of the world.

El Niño begins as a warming of the central equatorial Pacific. Over the ensuing six months to a year, this warming spreads eastward to the

Figure 32.5: An estimate of the total amount of power dissipated annually by Atlantic tropical cyclones (blue curve) and the "Southern Oscillation Index" (SOI), a measure of El Niño. When the SOI is small, El Niño is usually present (green curve). The Atlantic hurricane power has been scaled to have approximately the same range as the SOI.

▶

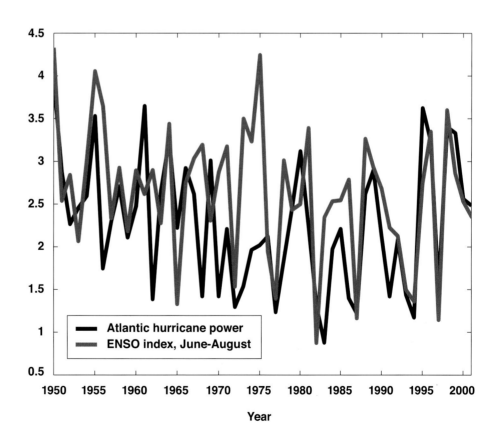

South American coast. Showers and thunderstorms that normally occur over the western tropical Pacific region shift eastward along with the ocean warmth, often causing severe flooding in the normally dry coastal regions of equatorial South America. At the same time, rainfall over Australia and the far western Pacific diminishes, causing drought and bush fires.

El Niño occurs sporadically, on average once every four years. Its arrival suppresses Atlantic hurricane activity, for reasons that remain somewhat mysterious. Figure 32.5 shows a measure of the annual power dissipated by Atlantic tropical cyclones together with a commonly used index of El Niño, called the "southern oscillation index," or SOI. (When the SOI is high, there is no El Niño; more negative values of the SOI indicate stronger El Niños.) The power dissipated by tropical cyclones is the sum over their lifetime of the surface wind speed cubed. Except for the period between about 1967 and 1975, the correlation between the two curves is quite good: when El Niño is present, there are fewer and/or weaker storms in the Atlantic. Although the reasons for this correlation are not well understood, forecasters can use it to predict Atlantic hurricane activity as much as a year in advance.

Bergeron also raised the question of how hurricane activity may have varied in the geologic past. Were there more or fewer hurricanes during the last ice age? Were they more intense or less? At various times in the earth's past, most recently about 50 million years ago, the climate was so warm that tropical plants grew well above the Arctic Circle and alligators roamed England. What were hurricanes like in these very warm climates?

Answers to these questions will prove more than just intellectual curiosities. Understanding how climate controls hurricane activity would allow us to predict how hurricanes might respond to future climate change, including anthropogenic global warming. If the earth, and the tropical oceans in particular, continue to warm up, will we be in for more hurricanes? And will they be more intense?

The answer to the second question appears to be yes: increasing the concentration of greenhouse gases in the atmosphere warms the tropical oceans and increases the potential intensity of hurricanes (see Chapter 12). The question is, by how much? We know that hurricane potential intensity increases by about 5 mph for each $1°F$ increase of tropical sea surface temperature. Climate models show that doubling the amount of CO_2 in the atmosphere should increase tropical ocean temperatures from $1°F$ to $4°F$, increasing potential intensity by 5 to 20 mph. Referring to Figure 12.1, we see that raising the potential intensity by, say, 10 percent will raise the intensity of all storms in the distribution by 10 percent. This implies that a storm that reaches a maximum wind speed of 100 mph in today's climate might achieve a maximum speed of 110 mph in an anthropogenically warmed world.

The possible effects of global warming on hurricane intensity is illustrated by a computer simulation of Hurricane Andrew (Chapter 31), using a simple computer model that includes coupling to the ocean (see Chapter 12). The model was first run using present-day ocean and atmosphere temperatures, and then rerun using conditions that might occur were the tropical oceans to warm up by about $4°F$. Figure 32.6 shows that the maximum wind speeds in Andrew might have been about 15 mph higher had it occurred in the warmer world.

Figure 32.6: Computer simulation of the maximum wind speed in Hurricane Andrew in the present climate (blue curve) and with hypothetical global warming that raises tropical sea surface temperatures by 4°F (red curve).

▶

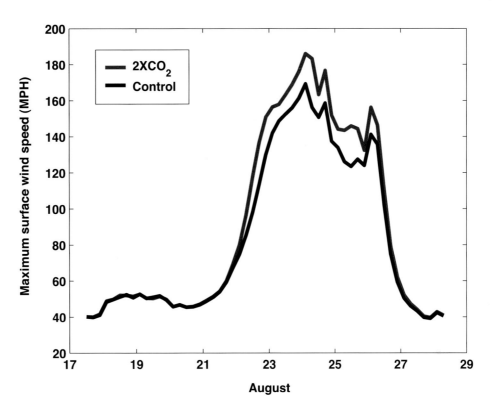

Although there is considerable evidence that hurricanes would be more intense in a warmer climate, the question of how frequently they might occur remains largely unanswered. Complex computer models of the global climate system now resolve the atmosphere well enough to produce storms that behave like tropical cyclones, but their resolution is still too coarse to capture the structure of the eye and eyewall, and consequently the modeled storms are too weak. But they do occur in about the right numbers and, mostly, in the right places. Such models have been used to predict how hurricane frequency would change in response to a doubling of atmospheric CO_2, but some models predict an increase in the number of storms while others predict a decrease. Science has not progressed far enough to allow us to say anything meaningful about how hurricane frequency might change in response to global warming. Quite possibly, the insights acquired from the new field of pale-

otempestology will lead to a better understanding of how hurricane activity might respond to global warming.

Hurricanes may have an important effect on earth's climate. Although they are too rare to have much direct effect on the atmosphere, they are efficient mixers of the upper tropical ocean (Chapter 12), and this mixing may have important effects on how the ocean carries heat away from the Tropics and toward the poles.

The ocean carries heat by moving warm water poleward near the surface, while deep in the ocean, cold water moves toward the equator; this overturning flow is known as the thermohaline circulation. Ice cores in Greenland and deep sea sediment cores suggest that a sharp slowing of the thermohaline circulation may have plunged the North Atlantic region back into an ice age just as it was recovering from the major episode of glaciation that ended about ten thousand years ago.

Remarkably, oceanographers still do not agree about what causes the thermohaline circulation. One thing seems clear: were it not for turbulence in the tropical oceans, the warm layer of water that today extends downward several hundred meters would collapse to a thin sheet only 50 m or so deep, and the overturning circulation would be very weak. Global hurricane activity may contribute to the tropical mixing that sustains the global thermohaline circulation. If this is so, then hurricanes may help stabilize tropical climate while destabilizing the climate at higher latitudes.

For example, lower global temperatures during ice ages would have sustained fewer and/or weaker hurricanes. The reduced mixing of the tropical upper ocean would have weakened the thermohaline circulation, transporting less heat away from the Tropics. This would keep the Tropics warm, while the poles, deprived of oceanic heat flow, would be even colder. Conversely, the hothouse climate that last existed about 50 million years ago may have been prone to exceptionally strong and frequent hurricanes, yielding a powerful thermohaline circulation. This might help explain why the poles were much warmer than at present, while the Tropics were only slightly warmer (Figure 32.7).

Bergeron's vision that we might someday understand how hurricanes depend on climate is only now beginning to be realized. Perhaps we will come to understand not only how hurricanes are influenced by climate but their role in regulating climate and in driving climate change.

Figure 32.7: Estimate of the sea surface temperature distribution during the early Eocene (about 50 million years ago; dashed lines show the range of estimates), compared to the present distribution (solid curve).

▶

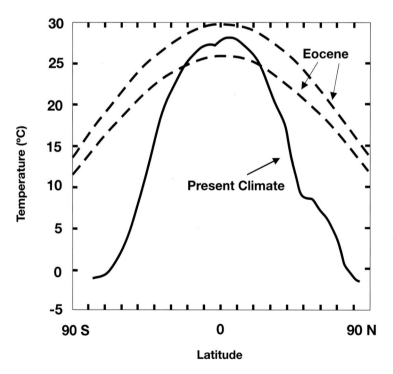

▲

"Storm over the Everglades at Sunset" seen from the Florida Keys. Photograph by Chris Kridler, August 2001.

And far across the silent sea
    in distant, dying light,
a long gone storm does yet bequeath
    a vivid missive through the night;
though,
silence

     —Anonymous

Epilogue

*D*ivine Wind goes to press during a year in which hurricanes have again exacted a terrible toll in human suffering. Jeanne, a mere tropical depression at the time, dumped stupendous quantities of rain in Haiti, killing more than two thousand and leaving countless more without homes. It then went on to become the fourth full hurricane to strike Florida in a period of a scant six weeks, a record for any one season, and pushing insurance payouts in 2004 to more than $22 billion, beating the old record set by Hurricane Andrew in 1992.

Halfway around the world, Typhoon Tokage became the seventh full typhoon and tenth tropical cyclone to hit Japan in 2004, also a new record, killing more than 55 and sending the year's tropical cyclone death toll to over 150.

The unusual number of landfalling hurricanes and typhoons predictably triggered a storm of speculation about whether tropical cyclones are responding to global warming. The general sense of alarm was compounded by the release of movies such as *The Day After Tomorrow*, which depicts terrifying superstorms set off in a climate driven crazy by human-induced global warming. In our anxious self-recrimination, are we so very different from the ancient Mayans, Caribs, and Tainos who believed they were being punished for their behavior?

If our actions are partly to blame, it is not for the reasons Hollywood would have us believe. While global warming may eventually have a noticeable effect on hurricane activity (Chapter 32), there is no evidence of a warming-induced trend in hurricanes so far, and while we focus on the distant future we ignore the peril at our doorstep. Scientists have been warning for some time that Atlantic hurricane activity is likely to return to levels recorded at various intervals in the past, as a consequence of poorly understood but well documented natural variability. A succession of directors of

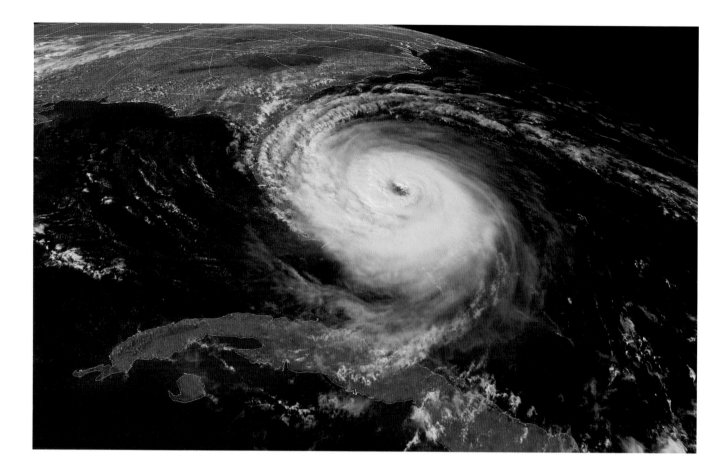

Hurricane Jeanne closes in
on Florida, September 2004

▲

the National Hurricane Center have for decades traveled the East and Gulf coasts of the United States, and testified to Congress, warning that the upswing in coastal population coupled with inadequate construction was setting us up for disaster. They have been largely ignored. Federal and state policies on such matters as flood insurance and disaster relief continue to encourage development in flood- and hurricane-prone regions, forcing all taxpayers to subsidize the risks undertaken by a few. In this respect, we could not differ more from the Mayans, who learned to build their cities well inland.

There is a silver lining on this most recent hurricane season: It happened to coincide with a major government-funded field experiment designed to understand the interactions between hurricanes and the oceans that fuel them. The large number of intense storms provided a windfall of valuable data that will help us better appreciate the workings of hurricanes, and also perhaps to better predict them.

Appendix I: Notable Tropical Cyclones

1274, November 19: Kublai Khan's first invasion of Japan is defeated by a typhoon, sustaining 13,000 casualties (Chapter 1).

1281, August 15: Kublai Khan's second and final attempt to invade Japan is again thwarted by a typhoon (Chapter 1).

1502, June 29: Columbus survives his first real encounter with a hurricane (Chapter 5).

1509, July 29: The city of Santo Domingo, established by Columbus, is almost completely destroyed by a hurricane.

1559, September 19: The first attempt by the Spanish to start a colony in Florida ends when a hurricane severely damages the fleet of Don Tristán de Luna y Arellano, who had been appointed to the task by the viceroy of New Spain, Luis de Velasco. The fleet was anchored in what is now called Pensacola Bay. The Spanish would not attempt to settle the Gulf coast of Florida again until 1693.

1565, September 19: Hurricane wrecks the French fleet off St. Augustine, allowing the Spanish to capture Fort Caroline and ending French claims to the region (Chapter 7).

1568, early September: A British fleet of slave traders, commanded by John Hawkins, is blown off course by a hurricane in the Gulf of Mexico and straggles into the Spanish-controlled port of San Juan de Ulloa, near Vera Cruz, Mexico, on September 16. After an initial truce, the Spaniards launched a surprise attack on the English, sinking all but two ships and killing many men. Hawkins commands one of the surviving ships; the other is captained by Francis Drake, whose bravery is brought to the attention of the queen of England, thereby launching his memorable career. Hawkins goes on to chair the English Naval Board and draws on his hurricane experience to redesign the English galleon, which would prove instrumental in establishing Britain's superior naval power.

1609, July 24–28: Fleet of English ships bound for Jamestown is wrecked on Bermuda by a hurricane. Some of the crew members remain on the island, establishing the first colony there. An account of the storm reaches England; it was almost certainly the basis for Shakespeare's *Tempest* (Chapter 9).

1635, August 25–26: A storm later named "the Great Colonial Hurricane" strikes New England. Many shipwrecks and several disasters occur during the storm, one of which would give birth to a favorite New England legend surrounding Thatcher's Island.

1640, September 11: Hurricane partially destroys a Dutch fleet poised to attack Havana, securing Spanish control of Cuba.

1666, August 4: Lord Willoughby, British governor of Barbados, loses most of his fleet of 17 ships and 2,000 troops in a hurricane near the Lesser Antilles. This allows the French to maintain control of Guadeloupe.

1667, August 27: First reported hurricane on the east coast of what was to become the United States strikes Jamestown, Virginia.

1715, July 29–31: Spanish gold fleet of 11 ships led by General Juan Esteban Ubilla, bound for Spain from Havana, is destroyed by hurricane off Florida. Only one ship survives.

1737, October 7: The Hooghly River Cyclone strikes India from the Bay of Bengal, killing over 300,000 and sinking more than 20,000 ships (Chapter 28).

1766, October 6: A hurricane devastates Martinique, ruining its sugarcane business. Among the families affected is that of one Joseph Tascher, who is financially ruined by the storm. Desperate about the well-being of his family, he sends his daughter, Rose, to France. Rose later marries Napoleon Bonaparte, who rechristens his new bride Josephine.

1772, August 31: Hurricane near Dominica sinks most ships in the area and destroys its sugarcane crop. Among the many islands experiencing extreme devastation is St. Croix, where the 17-year-old Alexander Hamilton writes an essay on the storm, impressing the island's Presbyterian clergyman, who has it printed in the island newspaper. The printing secures Hamilton's reputation as a youth of intellectual ability, and friends take up a collection to send Hamilton to the United States for a college education. The rest of his life is well known to history.

1775, September 12: The "Independence Hurricane" catches many fishing boats on the Grand Banks, killing 4,000 seamen, mostly from Britain and Ireland.

1776, September 6: Over 6,000 die when a major hurricane strikes Guadeloupe.

1780, October 2–7: The Savanna-la-Mar hurricane kills 3,000 in Jamaica, in Cuba, and at sea (Chapter 11).

1780, October 10–18: Hurricane kills more than 22,000 people in the eastern Caribbean region, the greatest loss of life in a single natural disaster in the Western world (Chapter 11).

1780, October 17–21: Solano's Hurricane kills 2,000 sailors in the eastern Gulf of Mexico (Chapter 11).

1782, September 16–17: More than 3,000 sailors perish in a central Atlantic hurricane.

1815, September 23: The Great September Gale, one of the most powerful hurricanes to strike New England, makes landfall at Long Island, New York, and then crosses Massachusetts and New Hampshire. It is the worst storm in nearly two centuries, equal in strength to the 1938 hurricane.

1821, September 1–4: Hurricane travels up U.S. East Coast, passing very close to New York City, which experiences a storm surge of at least

13 ft and suffers more damage than has ever been witnessed before in the city. This storm demonstrates that New York is vulnerable to hurricanes.

1831, August 10–11: A violent hurricane devastates Barbados. Death toll is estimated at 1,500 to 2,500 people. Damage survey by British engineer William Reid is instrumental in confirming William Redfield's theory that hurricanes are vortical (Chapter 2).

1844, October 4–5: Hurricane kills nearly 500 in Cuba. The damage done to Cuban agriculture by this storm, an earlier hurricane in 1842, and one two years later, in 1846, has lasting effects on Cuban economy and culture.

1856, August 10–11: Last Island, off the coast of Louisiana, is destroyed; almost all island inhabitants are lost. Story of this disaster becomes a regional legend and a novel, *Chita: A Memory of Last Island*, by Lafcadio Hearn.

1864, October 5: Cyclone and storm surge near the mouth of the Hooghly River destroys much of Calcutta and kills 50,000–70,000 (Chapter 28).

1867, October 29: Hurricane in the Virgin Islands and Puerto Rico kills 1,000. Pressure drops to 27.95" on St. Thomas.

1870, October 7–8: As many as 2,000 perish in a hurricane in Cuba.

1873, August 25–28: Hurricane destroys more than 1,200 ships in the Canadian Maritimes.

1876, October 31: A hurricane and storm surge strikes near the mouth of the Meghna River, in what is today part of Bangladesh, killing 100,000. Another 100,000 die of disease and starvation in the storm's aftermath (Chapter 28).

1881, October 8: Hurricane and storm surge destroys much of Haiphong, Vietnam, killing over 300,000.

1882, June 6: Arabian Sea cyclone kills more than 200,000 in and around Bombay.

1886, October 12: Indianola, Texas, destroyed, killing over 250. City is never rebuilt.

1893, August 27–28: Charleston, South Carolina, and many adjacent coastal islands destroyed, killing as many as 2,500.

1893, October 1–2: 1,800 killed in Mississippi delta, including New Orleans.

1897: More than 170,000 die as cyclone and storm surge sweep through East Pakistan (Chapter 28).

1899, August 8–19: Hurricane kills at least 3,000 from Puerto Rico to the Carolinas.

1900, September 8: Galveston, Texas, destroyed in worst natural calamity in U.S. history, with 8,000–12,000 dead (Chapter 13).

1909, August 28: As many as 2,000 die when a hurricane strikes northeastern Mexico.

1919, September 9–14: Storm kills at least 600 in Cuba, the Florida Keys, and Texas.

1923, September 1: Typhoon sweeps over Tokyo, Japan, and is followed by an earthquake that evening. Typhoon winds fan fires set by quake. 250,000 die in the double disaster.

1926, September 28: Intense hurricane strikes Miami, killing at least 300 and doing $1.7 billion (year 2000 dollars) in damage, the most expensive hurricane disaster in U.S. history until Hurricane Carol of 1954. It is estimated that if the same storm were to hit Miami today, it would cause $87 billion in damage (Chapter 15).

1926, October 20: More than 700 perish as hurricane sweeps across Cuba.

1928, September 16: After killing at least 1,500 people in the Caribbean, the great "Okeechobee Hurricane" washes much of the water out of the south side of Lake Okeechobee in Florida, killing as many as 3,000 in the second most lethal hurricane in U.S. history (Chapter 17).

1930, September 1–6: Hurricane kills between 2,000 and 8,000 in the Dominican Republic.

1931, September 6–10: Between 1,500 and 2,500 people are killed by a hurricane in Belize.

1932, November 9: Storm surge kills around 2,500 in Santa Cruz del Sur, Cuba.

1934, June 4–8: Between 2,000 and 3,000 die in a hurricane in Honduras and El Salvador.

1934, September 21: Typhoon kills more than 4,000 in Honshu, Japan.

1935, September 2: The "Labor Day Hurricane," the most intense ever to strike the United States, kills more than 400 in the Florida Keys (Chapter 19).

1935, October 25: More than 2,000 die in Haiti, mostly in the towns of Jérémie and Jacmel.

1938, September 21: More than 600 die in the fierce and unpredicted "Great New England Hurricane" (Chapter 21).

1939, September 25: Only known tropical cyclone of tropical-storm strength to strike California makes landfall near Long Beach, killing 45, mostly at sea.

1942, October 16: Over 35,000 perish south of Calcutta, India (Chapter 28).

1944, September 14–15: Hurricane along the U.S. eastern seaboard kills as many as 390.

1944, December 18: U.S. admiral William "Bull" Halsey on the Pacific theater of World War II loses 780 men, three destroyers, and more than a hundred aircraft in a violent typhoon (Chapter 23).

1951, August 17: Hurricane kills nearly 150 in Jamaica and another 50 in Mexico.

1953, September 25–27: Typhoon claims 250 in Japan and destroys a third of the industrial center of Nagoya.

1954: August 25–31: Hurricane Carol roars up the East Coast, killing 60 and doing more than $450 million in damages, the most expensive natural disaster in the United States until that time.

1954, October 5–18: Hurricane Hazel rampages through the Caribbean, then makes landfall in North Carolina, doing damage as far inland as Toronto, Canada (Chapter 16). It kills 800 in the Caribbean and another 200 in the United States and Canada.

1955, August 19: Hurricane Diane strikes New England, causing $5.5 billion in damage (in year 2000 dollars), the costliest hurricane in U.S. history until it is overtaken ten years later by Hurricane Betsy.

1959, September 26–27: Typhoon Vera kills 4,500 on Honshu Island, in Japan's deadliest typhoon.

1959, October 27: Rare hurricane on west coast of Mexico kills more than 2,000.

1961, October 27–31: Hurricane Hattie destroys Belize City, killing 270.

1963, May 28–29: Tropical cyclone in East Pakistan kills more than 22,000 (Chapter 28).

1963, September 30–October 8: Hurricane Flora kills more than 7,000 in Haiti and Cuba.

1965, September 7–10: Hurricane Betsy causes $8.5 billion in damages in Florida and Louisiana (in year 2000 dollars), the most expensive hurricane disaster in U.S. history until Agnes in 1972.

1965, May–June: Two tropical cyclones take more than 60,000 lives in East Pakistan (Chapter 28).

1969, August 17: Camille demolishes coastal Mississippi and Louisiana with recorded winds in excess of 90 m/s (200 mph). Storm surge reaches 9 m (30 ft) in places. More than 200 killed in spite of mass evacuation (Chapter 26).

1970, November 13: Upward of 500,000 killed by tropical cyclone in Bangladesh (Chapter 28).

1972, June 14–23: The remnants of Hurricane Agnes cause terrific flooding in the mid-Atlantic states, killing 122; leaving more than 330,000 homeless; and causing more than $8.6 billion in damage (adjusted for inflation to the year 2000). It is the most expensive natural disaster in U.S. history until Hugo in 1989.

1974, September 14–19: Flooding from Hurricane Fifi kills between 3,000 and 10,000 in Honduras.

1974, December 25: Cyclone Tracy, the deadliest in Australian history, demolishes the city of Darwin, killing 50 (Chapter 30).

1979, August 31: Hurricane David kills more than 1,200 in Santo Domingo.

1982, September 19: Tropical Storm Paul hits El Salvador and Guatemala; heavy rains kill at least 1,000 people.

1988, September 9–14: Hurricane Gilbert sets all-time minimum-pressure record for a North Atlantic hurricane, of 888 mb (26.23").

1989, September 17–21: Hurricane Hugo breaks record for damage in the United States, exceeded only by Hurricane Andrew, three years later. Losses amount to almost $10 billion, adjusted for inflation to the year 2000.

1991, October 28: Typhoon Thelma devastates the Philippines, and 6,000 people die by catastrophic events related to the storm, including dam failure, landslides, and extensive flash flooding. The most casualties occur on Leyte Island, where a 2.4 m (8 ft) storm surge strikes Ormoc, accounting for over 3,000 fatalities.

1992, August 24: Hurricane Andrew hits south Florida and becomes the most expensive natural disaster in U.S. history, causing $35 billion in damages, in U.S. dollars adjusted for inflation to the year 2000 (Chapter 31).

1998, October 22–November 5: Hurricane Mitch, after becoming the strongest October Atlantic hurricane on record, stalls over Honduras and Nicaragua, where rain-induced floods kill more than 11,000, making it the second-deadliest Atlantic hurricane in history (Chapter 24).

2004, August–September: A record four hurricanes strike Florida within two months, inflicting more than $22 billion in insured losses, exceeding the total insurance payment from Hurricane Andrew of 1992.

Appendix II: Hurricane Records

Most of these records were obtained from Chris Landsea's
Hurricane Frequently Asked Questions web site:
http://www.aoml.noaa.gov/hrd/weather_sub/faq.html

Highest wind speed: 85 m/s (190 mph)—Typhoon Tip, Northwest Pacific, October 1979;
Hurricane Camille, Gulf of Mexico, August 1969; Hurricane Allen, Caribbean, August 1980

Lowest central pressure: 870 mb (25.76" mercury)—Typhoon Tip, Northwest Pacific,
October 1979

Most rapid intensification: 44 m/s (98 mph) in one day—Typhoon Forrest, Northwest Pacific
Ocean, September 1983

Largest: Gale-force winds to a distance of 1,100 km (675 mi) from center—Typhoon Tip,
Northwest Pacific, October 1979

Smallest: Gale-force winds to a distance of 50 km (30 mi) from center—Tropical Cyclone Tracy;
Darwin, Australia; December 1974

Longest lasting: 31 days—Hurricane/Typhoon John, Northeast and Northwest Pacific,
August–September 1994

Highest storm surge: 13 m (42 ft)—Bathurst Bay Hurricane, Australia, 1899

Largest one-day rainfall: 180 cm (72 in)—Tropical Cyclone Denise, La Reunion Island,
January 1966

Closest to equator: 1.5° North Latitude—Typhoon Vamei, Indonesia, December 2001

Most deadly: >300,000 people—East Pakistan (Bangladesh), 1970

Most costly: $35 billion (in year 2000 U.S. dollars)—Hurricane Andrew; Bahamas, Florida,
and Louisiana; August 1992

Appendix III: Vortex on a Chip

Have a personal computer running Windows? Then in a few minutes, you can create your own custom hurricane, running an actual computer simulation. This computer model is a close cousin of one that is used to help make forecasts of hurricane intensity around the world. Just sign on to http://wind.mit.edu/~emanuel/modelftp.html and follow the instructions. You can specify the atmospheric environment through which the storm moves; change the ocean surface temperature, the humidity of the surrounding atmosphere, and other features; and allow the storm to be influenced by wind shear and other disturbances. You can also have the storm interact with the ocean, stirring up cold water from the depths and thereby weakening the storm. Warm or cold subsurface ocean eddies can also be added. The model storm can be programmed to make landfall, thereafter traveling over swampy, flat, hilly, or mountainous terrain. You will see on your screen graphics like the one pictured here:

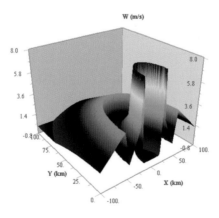

This shows a cross-section through a simulated storm. The height of the colored surface corresponds to the strength of the updraft velocity in the storm, with peak values here of about 7 m/s (15 mph), and the center of the storm is at lower right. This snapshot caught the simulated storm in the middle of an eyewall replacement, during which a new eyewall forms outside the existing one and contracts inward, eventually choking the old eyewall and taking its place. ("Cyclone" means "coil of a snake"; this one is a boa constrictor.) The image shown here is one of many graphics the model produces on the screen.

Sources and Further Reading

Adamson, H. C., and G. F. Kosco, 1967. *Halsey's Typhoons*. Crown Publishers, New York.

Allen, E. S., 1976. *A Wind to Shake the World: The Story of the 1938 Hurricane*. Little, Brown and Company, Boston.

Anthes, R. A., 1982. *Tropical cyclones, their evolution, structure and effects*. Meteorological Monographs. Boston, American Meteorological Society.

Barnes, J., 1998. *Florida's Hurricane History*. University of North Carolina Press, Chapel Hill.

Bender, M. A., I. Ginis, and Y. Y. Kurihara, 1993. Numerical simulations of tropical cyclone-ocean interaction with a high resolution coupled model. *Journal of Geophysical Research*, 98, 23245–63.

Benzoni, G., 1970. *History of the New World, Showing his travels in America, from A.D. 1541 to 1556, with some particulars of the island of Canary*. B. Franklin, New York.

Bergeron, T., 1954. The problem of tropical hurricanes. *Quarterly Journal of the Royal Meteorological Society*, 80, 131–64.

Bister, M., and K. A. Emanuel, 1997. The genesis of Hurricane Guillermo: TEXMEX analyses and a modeling study. *Monthly Weather Review*, 125, 2662–82.

Bluestein, H. B., and F. D. Marks, Jr., 1987. On the structure of the eyewall of Hurricane "Diana" (1984): comparison of radar and visual characteristics. *Monthly Weather Review*, 115, 2542–52.

Bosart, L. F., and G. M. Lackmann, 1995. Postlandfall tropical cyclone reintensification in a weakly baroclinic environment: a case study of Hurricane David (September 1979). *Monthly Weather Review*, 123, 3268–91.

Broccoli, A. J., and S. Manabe, 1990. Can existing climate models be used to study anthropogenic changes in tropical cyclone climate? *Geophysical Research Letters*, 17, 1917–20.

Bunbury, B., 1994. *Cyclone Tracy: Picking up the Pieces*. Fremantle Arts Centre Press, South Fremantle, Australia.

Burpee, R. W., 1972. The origin and structure of easterly waves in the lower troposphere of North Africa. *Journal of the Atmospheric Sciences*, 29, 77–90.

Calhoun, R. C., 1981. *Typhoon: The Other Enemy*. Naval Institute Press, Annapolis.

Carrier, J., 2001. *The Ship and the Storm: Hurricane Mitch and the Loss of the* Fantome. Harcourt, Inc., New York.

Chan, J. C.-L., and R. T. Williams, 1987. Analytical and numerical studies of the beta-effect in tropical cyclone motion. *J. Atmos. Sci.*, 44, 1257–64.

Conrad, J., 1938. *Typhoon and other stories*. Sun Dial Press, New York.

Corbosiero, K. L., and J. Molinari, 2003. The relationship between storm motion, vertical wind shear, and convective asymmetries in tropical cyclones. *Journal of the Atmospheric Sciences*, 60 (2), 366–76.

Davies, P., 2000. *Inside the Hurricane: Face to Face with Nature's Deadliest Storms*. Henry Holt and Company, New York.

DeMaria, M., 1985. Tropical cyclone motion in a nondivergent barotropic model. *Monthly Weather Review*, 137, 1199–1210.

———, 1996. The effect of vertical shear on tropical cyclone intensity change. *Journal of the Atmospheric Sciences*, 53, 2076–87.

Diaz, H. F., and R. S. Pulwarty, eds., 1997. *Hurricanes: Climate and Socioeconomic Impacts*. Springer, New York.

Douglas, M. S., 1958. *Hurricane*. Rinehart & Company, New York.

Drye, W., 2002. *Storm of the Century: The Labor Day Hurricane of 1935*. National Geographic, Washington, D.C.

Dunn, G. E., and B. I. Miller, 1960. *Atlantic Hurricanes*. Revised. Louisiana State University Press, New Orleans.

Elsberry, R. L., W. M. Frank, G. J. Holland, J. D. Jarrell, and R. L. Southern, 1987. *A Global View of Tropical Cyclones*. University of Chicago Press, Chicago.

Elsner, J. B., and A. B. Kara, 1999. *Hurricanes of the North Atlantic*. Oxford University Press, New York.

Emanuel, K. A., 1986. An air-sea interaction theory for tropical cyclones. Part I. *Journal of the Atmospheric Sciences*, 42, 1062–71.

———, 1987. The dependence of hurricane intensity on climate. *Nature*, 326, 483–85.

———, 1989. The finite-amplitude nature of tropical cyclogenesis. *Journal of the Atmospheric Sciences*, 46, 3431–56.

———, 1994. The physics of tropical cyclogenesis over the eastern Pacific. In *Tropical Cyclone Disasters*. J. Lighthill, Z. Zhemin, G. Holland, and K. Emanuel. Beijing, Peking University Press, 588.

———, 1995. The behavior of a simple hurricane model using a convective scheme based on subcloud-layer entropy equilibrium. *Journal of the Atmospheric Sciences*, 52, 3959–68.

———, 1997. Some aspects of hurricane inner-core dynamics and energetics. *Journal of the Atmospheric Sciences*, 54, 1014–26.

———, 1999. Thermodynamic control of hurricane intensity. *Nature*, 401, 665–69.

———, 2000. A statistical analysis of tropical cyclone intensity. *Monthly Weather Review*, 128, 1139–52.

———, 2003. Tropical cyclones. *Annual Review of Earth and Planetary Science*, 31, 75–104.

Emanuel, K., C. DesAutels, C. Holloway, and R. Korty, 2004. Environmental control of tropical cyclone intensity. *Journal of the Atmospheric Sciences*, 61, 843–58.

Fiorino, M., and R. L. Elsberry, 1989. Contributions to tropical cyclone motion by small, medium, and large scales in the initial vortex. *Monthly Weather Review*, 117, 721–27.

Frank, W. M., and E. A. Ritchie, 1999. Effects of environmental flow upon tropical cyclone structure. *Monthly Weather Review*, 127, 2044–61.

———, 2001. Effects of vertical wind shear on the intensity and structure of numerically simulated hurricanes. *Monthly Weather Review*, 129, 2249–69.

Goldenberg, S. B., and L. J. Shapiro, 1996. Physical mechanisms for the association of El Niño and West African rainfall with Atlantic major hurricane activity. *Journal of Climate*, 9, 1169–87.

Goldenberg, S. B., C. W. Landsea, A. M. Mestas-Nuñez, and W. M. Gray, 2001. The recent increase in Atlantic hurricane activity: Causes and implications. *Science*, 293, 474–79.

Gray, W. M., 1968. Global view of the origin of tropical disturbances and storms. *Monthly Weather Review*, 96, 669–700.

———, 1979. Hurricanes: Their formation, structure, and likely role in the tropical circulation. In *Meteorology over the tropical oceans*. D. B. Shaw, Roy. Meteor. Soc., 155–218.

Gray, W. M., C. Landsea, P. W. J. Mielke, and K. J. Berry, 1992. Predicting Atlantic seasonal hurricane activity 6–11 months in advance. *Weather and Forecasting*, 7, 440–55.

Hawkins, H. F., and D. T. Rubsam, 1968. Hurricane Hilda, 1964: II. Structure and budgets of the hurricane on October 1, 1964. *Monthly Weather Review*, 96, 617–36.

———, 1968. Hurricane Hilda, 1964. I: Genesis, as revealed by satellite photographs, conventional and aircraft data. *Monthly Weather Review*, 96, 428–52.

Hawkins, H. F., and S. M. Imbembo, 1976. The structure of a small, intense hurricane Inez, 1966. *Monthly Weather Review*, 104, 418–42.

Hearn, L., 2001. *Chita: A Memory of Last Island*. Pelican Publishing, Gretna, La.

Henderson-Sellers, A., H. Zhang, G. Berz, K. Emanuel, W. Gray, C. Landsea, G. Holland, J. Lighthill, S.-L. Shieh, P. Webster, and K. McGuffie, 1998. Tropical cyclones and global climate change: A post-IPCC assessment. *Bulletin of the American Meteorological Society*, 79, 19–38.

Holland, G., 1982. Tropical cyclone motion: environmental interaction plus a beta-effect. Colorado State University, Fort Collins, Colo.

Holland, G., and R. T. Merrill, 1984. On the dynamics of tropical cyclone structural changes. *Quarterly Journal of the Royal Meteorological Society*, 110, 723–45.

Hurston, Z. N., 1937. *Their Eyes Were Watching God*. Harper & Row, New York.

Jorgensen, D. P., E. J. Zipser, and M. A. LeMone, 1985. Vertical motions in intense hurricanes. *Journal of the Atmospheric Sciences*, 42, 839–56.

Khain, A., and I. Ginis, 1991. The mutual response of a moving tropical cyclone and the ocean. *Beiträge zur Physik de Atmosphäre*, 64, 125–41.

Kleinschmidt, E., Jr., 1951. Gundlagen einer Theorie des tropischen Zyklonen. *Archiv fur Meteorologie, Geophysik und Bioklimatologie*, Serie A, 4, 53–72.

Kurihara, Y., and A. M. Bender, 1982. Structure and analysis of the eye of a numerically simulated tropical cyclone. *Journal of the Meteorological Society of Japan*, 60, 381–95.

Kurihara, Y., R. E. Tuleya, and M. A. Bender, 1998. The GFDL hurricane prediction system and its performance in the 1995 hurricane season. *Monthly Weather Review*, 126, 1306–22.

Landsea, C. W., and W. M. Gray, 1992. The strong association between western Sahelian monsoon rainfall and intense Atlantic hurricanes. *Journal of Climate*, 5, 435–53.

Landsea, C. W., and J. and R. A. Pielke, 1998. Normalized hurricane damages in the United States, 1925–95. *Weather and Forecasting*, 13, 621–31.

Larson, E., 1999. *Isaac's Storm. A Man, a Time, and the Deadliest Hurricane in History*. Crown Publishers, New York.

Leipper, D. F., 1967. Observed ocean conditions and Hurricane Hilda, 1964. *Journal of the Atmospheric Sciences*, 24, 182–96.

Lighthill, J., A. Z. Zhemin, G. Holland, and K. Emanuel, 1993. *Tropical Cyclone Disasters*. Peking University Press, Beijing.

Liu, K.-B., and M. L. Fearn, 1993. Lake-sediment record of late Holocene hurricane activities from coastal Alabama. *Geology*, 21, 793–96.

Longshore, D., 1998. *Encyclopedia of Hurricanes, Typhoons, and Cyclones*. Facts on File, Inc., New York.

Lorenz, E. N., 1993. *The Essence of Chaos*. University of Washington Press, Seattle.

Ludlam, D. M., 1963. *Early American Hurricanes, 1492–1870*. American Meteorological Society, Boston.

Marks, F. D., Jr., and R. A. Houze, 1984. Airborne Doppler radar observations in Hurricane Debby. *Bulletin of the American Meteorological Society*, 65, 569–82.

_____, 1987. Inner core structure of Hurricane "Alicia" from airborne Doppler radar observations. *Journal of the Atmospheric Sciences*, 44, 1296–1317.

Merrill, R. T., 1984. Comparison of large and small tropical cyclones. *Monthly Weather Review*, 112, 1408–18.

_____, 1988. Environmental influences on hurricane intensification. *Journal of the Atmospheric Sciences*, 45, 1678–87.

Millás, J. C., 1968. *Hurricanes of the Caribbean and Adjacent Regions, 1492–1800*. Academy of the Arts and Sciences of the Americas, Miami.

Minsinger, W. E., 1988. *The 1938 Hurricane: An Historical and Pictorial Summary*. Greenhills Books, Randolph Center, Vt.

Montgomery, M. T., and B. F. Farrell, 1993. Tropical cyclone formation. *Journal of the Atmospheric Sciences*, 50, 285–310.

Montgomery, M. T., and R. Kallenback, 1997. A theory for vortex Rossby-waves and its application to spiral bands and intensity changes in hurricanes. *Quarterly Journal of the Royal Meteorological Society*, 123, 435–65.

Morison, S. E., 1971. *The European Discovery of America: The Northern Voyages, A.D. 500–1600*. Oxford University Press, New York.

Mowat, F., 2001. *The Serpent's Coil*. The Lyons Press, New York.

Möller, J. D., and M. T. Montgomery, 1999. Vortex Rossby waves and hurricane intensification in a barotropic model. *Journal of the Atmospheric Sciences*, 56, 1674–87.

———, 2000. Tropical cyclone evolution via potential vorticity anomalies in a three-dimensional balance model. *Journal of the Atmospheric Sciences*, 57, 3366–87.

Mykle, R., 2002. *Killer 'Cane: The Deadly Hurricane of 1928*. Cooper Square Press, New York.

Nelson, N. B., 1996. The wake of Hurricane Felix. *International Journal of Remote Sensing*, 17, 2893–95.

Nong, S., and K. Emanuel, 2003. Concentric eyewalls in hurricanes. *Quarterly Journal of the Royal Meteorological Society*, 129, 3323–38.

Nordhoff, C., and J. Hall, 1936. *The Hurricane*. Little, Brown and Company, Boston.

Ooyama, K., 1969. Numerical simulation of the life-cycle of tropical cyclones. *Journal of the Atmospheric Sciences*, 26, 3–40.

Ortiz, F., 1947. *El Huracan: Su Mitologia Y Sus Simbolos*. Fondo de Cultura Economica, Buenos Aires.

Palmén, E., 1948. On the formation and structure of tropical hurricanes. *Geophysica*, 3, 26–39.

———, 1958. Vertical circulation and release of kinetic energy during the development of Hurricane Hazel into an extratropical cyclone. *Tellus*, 10, 1–23.

Palmén, E., and A. C. W. Newton, 1969. *Atmospheric Circulation Systems*. Academic Press, New York.

Pérez, L. A., 2001. *Winds of Change: Hurricanes and the Transformation of Nineteenth-Century Cuba*. University of North Carolina Press, Chapel Hill.

Pielke, R. A. J., and R. A. S. Pielke, 1997. *Hurricanes: Their Nature and Impacts on Society*. John Wiley & Sons, New York.

Pielke, R. A., Jr., and C. W. Landsea, 1997. Normalized hurricane damages in the United States. *1925–1997. Weather and Forecasting*, 13, 351–61.

Potter, E. B., 1985. *Bull Halsey*. Naval Institute Press, Annapolis.

Powell, M. D., 1990. Boundary layer structure and dynamics in outer hurricane rainbands. Part I: Mesoscale rainfall and kinematic structure. *Monthly Weather Review*, 118, 891–917.

———, 1990. Boundary layer structure and dynamics in outer hurricane rainbands. Part II: Downdraft modification and mixed layer recovery. *Monthly Weather Review*, 118, 918–38.

Price, J. F., 1981. Upper ocean response to a hurricane. *Journal of Physical Oceanography*, 11, 153–75.

Price, J. F., and A. G. Z. F. T. B. Sanford, 1994. Forced stage response to a moving hurricane. *Journal of Physical Oceanography*, 24, 233–60.

Provenzo, E. F., Jr., and A. B. Provenzo, 2002. *In the Eye of Hurricane Andrew*. University Press of Florida, Gainesville.

Pudov, V. D., and Varfolomeev, N. Federov, 1979. Vertical structure of the wake of a typhoon in the upper ocean. *Okeanologiya*, 21, 142–46.

Reardon, L. F., 1926. *The Florida Hurricane and Disaster*. Miami Publishing Company, Miami.

Redfield, W. C., 1831. Remarks on the prevailing storms of the Atlantic Coast, of the North American States. *American Journal of Science and the Arts*, 20, 17–51.

Redfield, A. C., and A. R. Miller, 1957. Water levels accompanying Atlantic coast hurricanes. In *Interaction of Sea and Atmosphere*, Meteorological Monitor 2 (10): 1–23.

Reed, R. J., and E. E. Recker, 1971. Structure and properties of synoptic-scale wave disturbances in the equatorial western Pacific. *Journal of the Atmospheric Sciences*, 28, 1117–33.

Reid, W., 1849. The Progress of the Development of the Law of Storms and of the Variable Winds with the Practical Application of the Subject to Navigation. John Weale, London.

Riehl, H., and R. J. Shafer, 1944. The recurvature of tropical storms. *Journal of Meteorology*, 1, 42–54.

Riehl, H., 1948. On the formation of typhoons. *Journal of Meteorology*, 5, 247–64.

———, 1950. A model for hurricane formation. *Journal of Applied Physics*, 21, 917–25.

———, 1954. *Tropical Meteorology*. McGraw-Hill, New York.

Riehl, H., and J. S. Malkus, 1961. Some aspects of Hurricane Daisy, 1958. *Tellus*, 13, 181–213.

Riehl, H., 1963. Some relations between wind and thermal structure of steady state hurricanes. *Journal of the Atmospheric Sciences*, 20, 276–87.

———, 1975. Further studies on the origin of hurricanes.

Colorado State University, Fort Collins, Colo.

Ritchie, E. A., and G. J. Holland, 1993. On the interaction of tropical-cyclone scale vortices. II: Interacting vortex patches. *Quarterly Journal of the Royal Meteorological Society*, 119, 1363–97.

Rossby, C.-G., 1948. On displacement and intensity changes of atmospheric vortices. *Journal of Marine Research*, 7, 175–87.

Rotunno, R., and K. A. Emanuel, 1987. An air-sea interaction theory for tropical cyclones. Part II. *Journal of the Atmospheric Sciences*, 44, 542–61.

Samsury, C. E., and E. J. Zipser, 1995. Secondary wind maxima in hurricanes: Airflow and relationship to rainbands. *Monthly Weather Review*, 123, 3502–17.

Schade, L. R., and K. A. Emanuel, 1999. The ocean's effect on the intensity of tropical cyclones: Results from a simple coupled atmosphere-ocean model. *Journal of the Atmospheric Sciences*, 56, 642–51.

Shapiro, L. J., 1982. Hurricane climatic fluctuations. Part I: Patterns and cycles. *Monthly Weather Review*, 110, 1007–13.

———, 1982. Hurricane climatic fluctuations. Part II: Relation to large-scale circulation. *Monthly Weather Review*, 110, 1014–23.

———, 1983. The asymmetric boundary-layer flow under a translating hurricane. *Journal of the Atmospheric Sciences*, 40, 1984–98.

———, 1992. Hurricane vortex motion and evolution in a three-layer model. *Journal of the Atmospheric Sciences*, 49, 140–53.

Shapiro, L. J., and S. B. Goldenberg, 1998. Atlantic sea surface temperatures and tropical cyclone formation. *Journal of Climate*, 6, 677–99.

Shay, L. K., 1999. Upper ocean response to tropical cyclones. Rosenstiel School of Marine and Atmospheric Science Report 99-003, Miami.

Shay, L. K., G. J. Goni, and P. G. Black, 2000. Effects of a warm oceanic feature on Hurricane Opal. *Monthly Weather Review*, 128, 1366–83.

Sheets, B., and J. Williams, 2001. *Hurricane Watch: Forecasting the Deadliest Storms on Earth*. Vintage Books, New York.

Simpson, R. H., and H. Riehl, 1958. Mid-tropospheric ventilation as a constraint on hurricane development and maintenance. Technical Conference on Hurricanes, American Meteorological Society.

———, 1981. *The Hurricane and its Impact*. Basil Blackwell Publisher, Oxford.

Simpson, J., E. A. Ritchie, G. Holland, J. Halverson, and S. Stewart, 1997. Mesoscale interactions in tropical cyclone genesis. *Monthly Weather Review*, 125, 2643–61.

Simpson, R. H., ed., 2003. *Hurricane! Coping with Disaster*. American Geophysical Union, Washington, D.C.

Smith, R. K., 1980. Tropical cyclone eye dynamics. *Journal of the Atmospheric Sciences*, 37, 1227–32.

———, 1991. An analytic theory of tropical-cyclone motion in a barotropic shear flow. *Quarterly Journal of the Royal Meteorological Society*, 117, 685–714.

Smith, R. K., and W. Ulrich, 1990. An analytical theory of tropical cyclone motion using a barotropic model. *Journal of the Atmospheric Sciences*, 47, 1973–86.

Smith, R. K., and H. Weber, 1993. An extended analytic theory of tropical-cyclone motion in a barotropic shear flow. *Quarterly Journal of the Royal Meteorological Society*, 119, 1149–66.

Solá, E. M., 1995. *Historia de los Huracanes en Puerto Rico*. First Book Publishing, San Juan.

Southern, R. L., 1979. The global socio-economic impact of tropical cyclones. *Australian Meteorological Magazine*, 27, 175–95.

Standiford, L., 2002. *Last Train to Paradise*. Three Rivers Press, New York.

Sutyrin, G. G., and A. P. Khain, 1984. Effect of the ocean-atmosphere interaction on the intensity of a moving tropical cyclone. *Izvestiya, Atmospheric and Oceanic Physica*, 20, 697–703.

Swanson, R. L., 1976. Tides. In *Marine Ecosystem Analysis*. New York Bight Atlas, Monograph #4, January 1976. New York Sea Grant Institute, Albany.

Tannehill, I. R., 1952. *Hurricanes: Their Nature and History*. Princeton University Press, Princeton.

———, 1955. *The Hurricane Hunters*. Dodd, Mead & Company, New York.

Toomey, D., 2002. Storm Chasers: The Hurricane Hunters and their Fateful Flight into Hurricane Janet. W. W. Norton & Company, New York.

Tucker, T., 1982. *Beware the Hurricane!* Island Press Limited, Bermuda.

Viñes, B., 1885. *Practical hints in regard to West Indian hurricanes*. U.S. Government Printing Office, Washington, D.C.

Walter, H. J., 1992. The Wind Chasers: The History of the U. S. Navy's Atlantic Fleet "Hurricane Hunters." Taylor Publishing, Dallas.

Wang, Y., and G. Holland, 1996. Tropical cyclone motion and evolution in vertical shear. *Journal of the Atmospheric Sciences*, 53, 3313–32.

Watson, B., ed., 1990. *New England's Disastrous Weather*. Yankee Books, Camden, Maine.

Weatherford, C. L., and W. M. Gray, 1988. Typhoon structure as revealed by aircraft reconnaissance. Part I: Data analysis and climatology. *Monthly Weather Review*, 116, 1032–43.

Will, L. E., 1961. *Okeechobee Hurricane and the Hoover Dike*. Great Outdoors Association, St. Petersburg.

Williams, J. M., and I. W. Duedall, 2002. *Florida Hurricanes and Tropical Storms, 1871–2001*. University Press of Florida, Gainesville.

Willoughby, H. E., J. A. Clos, and M. G. Shoreibah, 1982. Concentric eyes, secondary wind maxima, and the evolution of the hurricane vortex. *Journal of the Atmospheric Sciences*, 39, 395–411.

Willoughby, H. E., F. D. Marks, Jr., and R. J. Feinberg, 1984. Stationary and moving convective bands in hurricanes. *Journal of the Atmospheric Sciences*, 41, 3189–3211.

Willoughby, H. E., H.-L. Jin, S. J. Lord, and J. M. Piotrowicz, 1984. Hurricane structure and evolution as simulated by an axisymmetric nonhydrostatic numerical model. *Journal of the Atmospheric Sciences*, 41, 1169–86.

Willoughby, H. E., and P. G. Black, 1996. Hurricane Andrew in Florida: Dynamics of a disaster. *Bulletin of the American Meteorological Society*, 77, 643–52.

Wu, C.-C., and K. A. Emanuel, 1993. Interaction of a baroclinic vortex with background shear: Application to hurricane movement. *Journal of the Atmospheric Sciences*, 50, 62–76.

_____, 1995. Potential vorticity diagnostics of hurricane movement. Part I: A case study of hurricane Bob (1991). *Monthly Weather Review*, 122, 69–92.

_____, 1995. Potential vorticity diagnostics of hurricane movement. Part II: Tropical Storm Ana (1991) and Hurricane Andrew (1992). *Monthly Weather Review*, 122, 93–109.

Yanai, M., 1964. Formation of tropical cyclones. *Reviews of Geophysics*, 2, 367–414.

Zehr, R. M., 1991. Observational investigation of tropical cyclogenesis in the western North Pacific. Preprints of the 19th Conference on Hurricanes and Tropical Meteorology, American Meteorological Society, Miami.

_____, 2002. Vertical wind shear characteristics with Atlantic hurricanes during 2001. 25th Conference on Hurricanes and Tropical Meteorology, American Meteorological Society, Ft. Lauderdale.

Credits

Unless otherwise noted, illustrations are by the author.

Index